GRASSLAND OF CHINA

Edited by Yan Weihong, Liu Lei, Xu Zhu

China Agricultural Science and Technology Press

图书在版编目（CIP）数据

中国的草原 = Grassland of China：英文 / 徐柱，雍世鹏，阎贵兴等主编；闫伟红，刘磊，徐柱等译. — 北京：中国农业科学技术出版社，2016.12
　　ISBN 978-7-5116-2798-8

Ⅰ.①中… Ⅱ.①徐…②闫… Ⅲ.①草原—概况—中国—英文 Ⅳ.①S812

中国版本图书馆 CIP 数据核字（2016）第 252733 号

Funded project: National Natural Science Foundation of China "Epigenetic diversity of *Medicago falcata* L. at different latitudes and DNA analysis of epigenetic changes under low temperature stress (31302017)"

责任编辑　李冠桥
责任校对　贾海霞

出 版 者	中国农业科学技术出版社 北京市中关村南大街 12 号　邮编：100081
电　　话	（010）82109705（编辑室）（010）82109704（发行部） （010）82109709（读者服务部）
传　　真	（010）82106625
网　　址	http://www.castp.cn
经 销 者	各地新华书店
印 刷 者	北京科信印刷有限公司
开　　本	787 mm × 1092 mm　1/16
印　　张	11.75
字　　数	278 千字
版　　次	2016 年 12 月第 1 版　2016 年 12 月第 1 次印刷
定　　价	50.00 元

◆ 版权所有·侵权必究 ◆

EDITORIAL BOARD

Translator

Yan Weihong, Liu Lei, Xu Zhu, Chen Libo

Proofreader

Qi Juan, Shan Guilian, Zheng Yang, Shi Wengui

Contributor(from left to right)

Yan Weihong, Liu Lei, Xu Zhu, Chen Libo,

Qi Juan, Shan Guilian, Zheng Yang, Shi Wengui,

Ning Fa, Ma Yubao, Wang Kai, Tian Qingsong, Wu Hongxin,

Zhang Xiaoqing, Yong Shipeng, Yan Guixing, Xing Lianlian,

Xie Jihong, Meng Qingwei, Zhang Jingran, Han Huizhi

Photo provided by

Xu Zhu, Yong Shipeng, Xing Lianlian, Shan Guilian, Bartel,

Li Xin, Li Chunrong, Li Xiaohui, Pan Yanqiu, Shi Wengui,

Liu Yu, Wang Shunli, Yan Zhijian, Yang Guisheng, Yu Yonggang,

Luo Yinggang, J.Marc Foggin, J.M.Suttic

Contents

Preface

Chapter 1 Overview of Grassland
2   / What is Grassland
2   / Where is Grassland
7   / Fountainhead of Grassland
9   / Ecological Functions of Grassland
10 / Glance of Grasslands of the World

Chapter 2 Grassland Environment
18 / Physiognomy
21 / Climate
23 / Soil
25 / Hydrology

Chapter 3 Grassland Vegetation
30 / Species Composition of Grassland Vegetation
33 / Floristic Geographical Elements of Grassland Vegetation
33 / Life Form of Grassland Vegetation
39 / Ecological Group of Grassland Vegetation
41 / Dynamics of Grassland Vegetation
43 / Main Formations of Grassland Vegetation

Chapter 4 Grassland Animal
60 / The ecological position of the grassland animals
60 / Coevolution of grassland animals and grassland environment
67 / Distribution of Grassland Animals in the World
73 / Grassland Animals of China

Chapter 5　Grassland Eco-culture
96　/ Walk into the grassland, walk into a herdsman's home
100　/ Unique Charm of Grassland
108　/ Advance with the times, march toward modernization

Chapter 6　Grassland Eco-tourism
112　/ Welcome to grassland
120　/ Relics on the grassland

Chapter 7　Landscape Regions of China's Grassland
128　/ Bunchgrass Grassland Ecoregion in Inner Mongolia Plateau
130　/ Forest Steppe Ecoregion in the Northeast Plain
132　/ Warm-temperate Grassland Ecoregion in Loess Plateau
134　/ Alpine Steppe Ecoregion of Qinghai-Tibet Plateau
137　/ Mountainous Bunch Grass, Dwarf Forb Desert Steppe Ecoregion in Xinjiang Altai

Chapter 8　Sustainable Utilization and Management of Grassland Resources
144　/ Historical Experience and Lessons of Exploitation and Utilization of Grasslands
147　/ Current Situation of Grassland Degradation and Comprehensive Prevention Countermeasures

151　/ Ecological Principles and Technical Strategy of Reasonable Utilization of Grassland
157　/ Grading Criterion of Grassland Degradation
159　/ Management of Grazing Land on the Grassland
164　/ Management of Mowing Pasture of Grassland
167　/ Management of Resources of Medicinal Plants on Grassland
169　/ Systemically Developing Grassland Dynamic Monitoring and Realizing the Sustainable Utilization of Grassland Resources

173　/ Postscript
175　/ References

Preface

The Old to the Chile Prairie
Below the Yinshan Mountains lies the Chile Prairie.
Over the earth hangs the sky like a huge yurt.
Between the vast sky and the boundless earth,
Flocks and herds appear as grass bends to wind.

Blue sky, white clouds, green grass, galloping horses and snow-white flock of sheep or goats, make it a fascinating place.

The grassland is a very important part of continental ecosystem. It forms an independent ecological niche with the forest and desert. Its production, distribution as well as growth and decline have certain laws of nature.

The grassland looks like the bright and colorful natural scroll, which integrates heaven, earth and mankind. Forest steppe, meadow steppe, typical steppe, desert steppe, diversified vegetation composition and ecological environment, all of which form the amazing landscape in various forms. The grassland is the vegetable kingdom in the semi-arid and arid areas, and is paradise and home to many wild animals and pastoralists.

Grassland is the well-carefully crafted product of nature, it has a long history. The grassland is the cradle of Chinese ancient nomadic nationalities such as Huns, Xianbei, Chile, Turkic, Khitan, Jurchen, Mongolian, Kazak, Uygur and Tibetan and other nomad's growth. Theygather and start their legend life in this vast prairie, pitching camps, riding horses, staging a thrilling historical drama, and leaving many unforgettable memories.

This book will help you understand the distribution pattern and general picture of grasslands of both China and the world, special ecological geographical conditions and vegetation characteristics of different grasslands and evolutionary changes of grassland vegetation, as well as some knowledge on rational utilization and scientific management of grasslands, and get a further insight into the little-known living habits and interesting news of animals living on grasslands. Furthermore, the book presents the true slice of production and life of herdsmen, unique grassland culture and national customs from various angles of view. Readers can appreciate the natural beauty of grassland, the beauty of soul of people who live in grasslands and harmonious beauty of the unity of heaven, earth and men.

This book is only the tip of an iceberg of grassland due to space constraints. Wish the sky bluer and grass greener.

The beautiful grassland is my homeland, and my forever love.

Chapter 1
Overview of Grassland

I've got an Appointment with the Grassland
Written by Yang Yanlei, composed by Siqin Zhaoketu

Always want to see your smiling face
Always want to hear your ringing voice
Always want to live in your yurt
Always want to raise your wine goblet
I've got an appointment with the grassland
To search our common root of the moland
Now that on the way back home I am coming
Into the sunshine to embrace the spring
Seeing your smiling face so pure
And feeling your sweet voice such a lure
Living in your yurt I find it so warming
Tasting your fermented milk, so intoxicating
I've got an appointment with the grassland
To warship the God in heart as we understand
Now that I am stepping into the home's gate
I can't help my running tears and a pounding heart
I used to look at you afar
I used to hug you in dream
I used to pray for you silently
I used to be attracted by you dearly
I've got an appointment with the grassland
To pour our love and speak our mind
Now that I am in your breast-grassland
Let's make the meeting eternal by holding the land

What is Grassland

Grassland may be defined as ground covered by vegetation dominated by herbaceous plants and shrub, also known as rangeland.

Grassland is an important part of terrestrial ecosystem and forms the green vegetation together with forest and other terrestrial ecosystem and becomes the integrated creature's living environment and renewable resources. Due to different geographic location, grassland's ecological environment has many tropisms and formed a variety of types.

The graph of ecological relationships among grassland types

Among which:

Typical steppe: xerophytic herbaceous community adapted to semi-drought habitat.

Meadow steppe: mesoxerophytes herbaceous community adapted to semi-humid habitat.

Desert steppe: xerophytic herbaceous community adapted to arid habitat.

Cold steppe: cold-xerophytic community adapted to highland habitat where it is characterized by rarefied air, high altitude, strong solar radiation and low temperature. It is also called ice and snow grassland or barren grassland.

Salt steppe: plant community adapted to the seasonal soil moisture and the salinity fluctuation habitat, also known as salt steppe.

To sum up, we can form such an idea that the grassland is one kind of the unusual ecosystem type in varied grassland types. The book is confined to discuss temperate grasslands that are broadly distributed in Northeast, North China, Northwest of China, and the cold steppe of Qinghai-Tibet Plateaus as well salt damp grassland of Northeast.

Where is Grassland

Grassland is a kind of vegetation type and a unique physical geography where the grassland biome

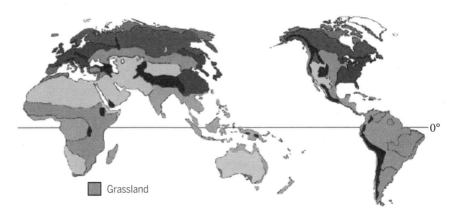

Distribution Diagram of Grassland of World (quoted from World Grassland)

is often something in between desert and forest.

When spreading map of the world vegetation, we can clearly see the renowned Eurasian grassland, which looks like a green stripe wide or narrow that located in the mid-latitude region of northern hemisphere between desert and forest.

Grassland of China seems as a string of bright pearls enchased on the east of Eurasian grassland zone. It is not only bright-coloured dazzling, but also attractive and enchanting, showing particular geographic location and various ecotypes of grasslands.

The grassland of China starts from the northeast Songliao Plain and traverses to Inner Mongolia Plateau, Erdos Plateau, Loess Plateau and ends in the Qinghai-Tibet Plateau. It stretches fromnortheast to southwest for about 4,500 km. Affected by southeast monsoon and Mongolian high-pressure and restricted by landform conditions, the distribution of east-west Eurasian grassland zone transfers largely from northeast to southwest in China, making China's grasslands adjacent to northern coniferous forest, extratropical mixed deciduous forest in East Asia, warm-temperate deciduous broad-leaf forest, subtropical evergreen broad-leaf forest, tropical monsoon forest, temperate deserts and cold desert. Such boundary relations with grasslands around the world and ecological effect are extremely rare. This special ecogeographical pattern has different influences on grassland landscape. And also, under the unique climate conditions, grassland has another effect on neighborhood vegetation. In the long geological history process, the buffer transition zone such as "forest-grassland" and "desert-grassland" on the peripheral area of the center area of the typical steppe have been formed.

Unique geographic location has shaped many different grassland ecozone types. Widely distributed in China, the grassland stretches 22 latitude degrees from south to north and 44 longitude degrees from east to west with elevations ranging from 100 m to above 5,000 m. On vertical direction, the altitude hoists from 100 m to 5,000 m. Within such an extensive geographical space, quantity of heat reduces gradually from north to south and from low altitude to high altitude and moisture content drops off gradually along the coast to inland, thus causing significant difference of zonality of China's grassland vegetation.

1. Rules of zonality of heat of grassland vegetation along the latitudinal direction (from south to north)

Heat difference of earth's surface caused by different solar incident angles is the leading factor for latitudinal zonal difference. In inland region of China, southward moving of latitude and the elevation of terrain happened at the same time, which greatly accelerated the southward extension of temperate grassland. It can be marked off three grassland thermal ecotypes distinctly from north to south: mid-temperate grassland, warm-temperate grassland andalpine steppe.

Mid-temperate grassland is located in Inner Mongolia Highland, Northeast Songnen Plain and Xiliao River Plain of the northern Yinshan Mountains.

The heat conditions of such grassland are moderate with an annual average temperature of -3~4.7 ℃ and accumulative temperature of ≥ 10 ℃ is 1,664~2,625 ℃.

Warm-temperate grassland is located in Erdos and Loess Plateau of the southern Yinshan Mountain and extends westward to Qinghai Lake. The thermal condition is the highest among the temperate grassland with an annual average temperature of 4.5~9 ℃. The annual accumulative temperature ≥10 ℃ is 2,370~3,300 ℃.

Alpine steppe is distributed in Qinghai-Tibet Plateau, which is the highest plateau and the most unique grassland of the world. The thermal condition is the lowest among temperate grasslands. The annual average temperature is -4~4 ℃. The annual average accumulative temperature ≥10 ℃ is lower than 1,000 ℃. The significant differences between alpine steppe and the former two type temperature grasslands are strong solar radiation, lower accumulative temperature, shorter growth season, larger difference in temperature and evident alternative freezing and thawing.

2. Rules of zonality of humidity of grassland vegetation along the longitude direction (from west to east) along longitude direction

Except for the territorial difference of hot conditions, the atmospheric precipitation of grassland decreases gradually from southeast to northwest and the dryness increases gradually because of multiple mutual influences such as southeastern monsoon, high pressure of Mongolia and Qinghai-Tibet plateaus, which further led to the differentiation of quantity of heat and moisture content and also thus formed three types of grasslands: meadow steppe, typical steppe and desert steppe.

The east and south sides of the forest region have an annual precipitation of 350~450 millimeters and moisture content of 0.7~0.9. It belongs to semi-humid climate scattered with meadow steppe, low mountains and hills, as well as island or cloddy forests.

The area close to the desert has an annual rainfall of 150~250 millimeters, and moisture content of 0.12~0.24, and belongs to arid climate. It has sparse and low desert steppe.

The annual precipitation of central part of the grassland fluctuates between 250~350 millimeters and humidity is 0.3~0.6. It has a semi-arid climate, which has formed typical steppe, also known as genuine grassland (Table 1-1).

3. The vertical distribution law of uplands vegetation of grassland regions

Except above-mentioned horizontal distribution of grassland, the vegetation shows vertical distribution with the changes of altitude.

The Daxinganling Mountain is the uppermost upland in mid-temperate zone grassland in China. Its north section is in the coniferous forest area of cold temperate zone, and south section stretches into the grassland areas. The terrain of Daxinganling Mountain slopes gently with an average elevation of 1,700 meters. Its west side is Inner Mongolia Plateau with altitude of about 650~700 m, where typical steppe is located; when height rises to 700~800 m, birch forest steppe appears

Table 1-1 The Basic Ecotypes and Brief Table of Grassland of China

Type	Mid-temperate Grassland			Warm-temperate Grassland			Alpine Steppe		
Subtype / Index	Meadow Steppe	Typical Steppe	Desert Steppe	Meadow Steppe	Typical Steppe	Desert Steppe	Meadow Steppe	Typical Steppe	Desert Steppe
Mean annual temperate (0°C)	-3~+3.1	-2.3~+4.5	+2.6~+4.7	+6.8~+7.5	+4.5~+7.8	+6.0~+9.0	-2.0~+4.0	-2.0~0.0	-4.0~-2.0
≥10°C Accumulative Temperature	1,664~1,693	1,768~2,385	2,023~2,625	3,033~3,214	2,370~3,200	2,623~3,300	>500~<1,000	500~1,000	<500
Annual precipitation (mm)	357~426	218~445	150~280	416~558	330~477	200~302	300~400	150~300	75~150
Humidity (K)	0.70~0.90	0.30~0.60	0.12~0.27	0.40~0.50	0.30~0.50	0.20~0.24	0.40~0.59	0.32~0.42	0.18~0.28
Distributing Area	Songnen Plain, Southern part of Daxinganling Mountain, East of Inner Mongolia, Yinshan Mountain, Tianshan, Altai Mountains	Xiliao river plain, Central Inner Mongolian Plateau	Wulanchabu Plateau	East of Loess Plateau	Central Loess Plateau, East of Erdos Plateau	Northwest of Loess Plateau, Midwest of Erdos Plateau	Qiangtang Plateau-Southeast, Qilian Mountain, Tianshan Mountain, Altai Mountain	Qiangtang Plateau Mid-part, Qilian Mountain, Tianshan Mountain, Altai Mountain	North of Qiangtang Plateau
Altitude (m)	150~1,200	450~1,100	900~1,500	500~1,000	900~1,800	1,100~1,800	3,600~5,000	4,500~4,700	4,200~5,020
Soil	Chernozem	Chestnut soil	Brown soil	Loessial soil	Chestnut-brown soil	Sierozem	Alpine Meadow	Alpine Grassland	Alpine Desert steppe
Plant Eco-group	Xerophytic grass, mesophyte, xerophytic-mesophyte forb	Typical xerophytic grass	Strong xerophytic low grass, short subshrub	Thermophilic xerophytic mesophyte, xerophytic, mesophyte forb, shrub	Thermophilic typical xerophytic grass	Thermophilic strong xerophytic grass shrub	Cold-xerophytic grass *Carex*, *Kobresia myosuroides* forbs	Strong cold-xerophytic grass *Carex*	Strong cold-xerophytic grass, *Carex*, dawf sub shrub
Crop Composition	Spring wheat, Naked oats, Potato, Rape	—	—	Winter wheat, Millet, Grain sorghum, Maize, Sweet potato, Peanut	Spring wheat, Potato	—	Highland barley	—	—

Quoted from *Grasslands of China*, Zhang Minghua

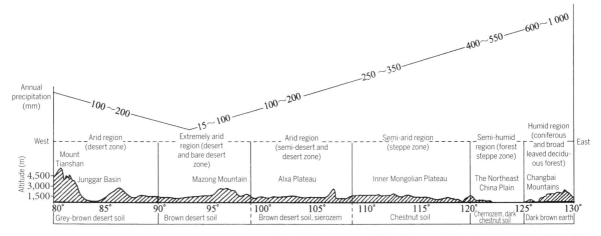

Longitudinal Change of Horizontal Distribution of Vegetation in China's Temperate and Warm-Temperate Zones (Latitude 40°~45°N) & Its Relationship with Meteoric Waters and Soil.doc

on hills before mountain; pure forest of white birch appears above 800 m; when altitude reaches to 900~950 m, it has Xingan larch forest. On the east side of Daxinganling Mountain lies the Songliao Plain, its altitude is below 300 m. *Prunus armeniaca—Stipa grandis* shrub grassland is located on the baseband of the front mountain. When the height arrives at 350~400 m or so, *Stipa baicalensis*, meadow steppe of *Filifolium sibiricum* and island *Quercus mongolicus* come into sight. When the altitude reaches 450~900 m, *Betula dahurica* and *Quercus mongolicus* are in a dominant position. When the altitude is above 900 m, it is replaced by Xing' an larch forest.

Grassland belt of warm temperature zone have many uplands, such as Yinshan Mountain, Helan Mountain and Liupan Mountain, etc. The vegetation baseband of these hills are temperate type typical steppe or desert steppe. With the rising of altitude, summer green broad-leaved forest, subalpine scrub, sub-alpine meadow or sub-alpine krummholz, alpine scrub and alpine meadow appear in order.

The alpine steppe of Qinghai-Tibet Plateau sits many lofty mountains such as Gangdise Mountain and

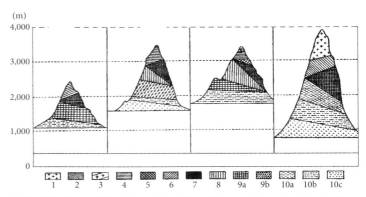

Vertical Spectrum of Mountainous Vegetation of Temperate Grassland (quoted from Vegetation of China)

1. Alpine sparse vegetation zone   2. Alpine steppe zone   3. Alpine meadow zone   4. Alpine shrub zone   5. Subalpine elfin wood zone   6. Subalpine meadow, bushwood and upland meadow zone   7. Mountainous coniferous forest zone   8. Subalpine shrub zone   9. Broadleaved deciduous forest zone   a. oak forest   b. Xerophytic bushwood and elm woodland   10. Steppe zone   a. Forest steppe subzone   b. Typical steppe subzone   c. Desert steppe subzone

Nyainqentanglha Mountain with an average altitude of above 6,000 m. A group of unique vegetation type of high mountains, such as alpine sparse vegetation, alpine snow and ice vegetation has been formed. According to scientific evidence, the vegetation of Qinghai-Tibet Plateau is not homogeneous, which has distinct zonal differentiation. Approximately, along with the terrain rising gradually from southeast to northwest, it distributes mountain forest zone (the evergreen broad-leaf forest, cold-temperate coniferous forest) →alpine bush, alpine meadow→ alpine grassland zone (the valley of lower altitude is temperate grassland)→alpine desert zone (the drought dale and valley slope of lower height is temperature mountain desert) in order. The relationships among highland surface meadow, grassland and desert zone are really not vertical gradient, but are highland zonal series that along horizontal direction happened in highland surface, a presentation of zonal characteristics of highlands.

Fountainhead of Grassland

Grassland is the evolutionary product of history in certain period of time. It is well known that angiosperm started to thrive at Cretaceous period on earth (70,000,000 years ago), and has developed into one of the biggest plant groups gradually. Among of these, Poaceae (the grass family) has the closest relationship with grassland. It began to differentiate after the Tertiary Metaphase (10,000,000 years ago) and distributed to the whole world very quickly. Now there are about 4,500 species of the grass family, some of which are the constructive species of grassland, including *Stipa* known as "the king of grassland". Not long ago, *Stipa* fossil was discovered in Miocene stratum of North America (7,000,000 years ago) and ancient horse fossil of Eurasian grassland animal was found at the same period. It was explained that grassland animals should appear on earth no later than seven million years ago. At the same time, the ancient geographical data confirmed that as the global temperature came down during the late Oligocene, the dry and cold climate created conditions for the appearance and expanding of grasslands.

The earth's crust had experienced great changes from Miocene (23,000,000~5,300,000 years ago) to the middle period of Quaternaryin China. Qinghai-Tibet Plateau arched in western China, and the eastern coastal areas also rose to form a series of mountains, and Inner Mongolian Plateau and Loess Plateau were formed simultaneously. These changes obstructed moist flow of North Atlantic Ocean from going to the east and wet flow of Indian Ocean from flowing to the north. Meanwhile, it also greatly weakened the westward extension of moist air mass of Pacific Ocean, which made the northern China become increasing drought, contributing to the formation of arid and semiarid areas of China. Moreover, according to paleomagnetic data, from late Eocene to this time, the geographical location of China has moved 10~13 degrees latitude northward, making the air temperature gradually decline. Arid area of China took on sparse grassland landscape in the Miocene period, and formed segments of steppe community only when mountains came into being. At the end of Miocene, as the weather became colder, the grassland communities that have formed on upland dropped to the flat ground and moved eastward. The alternation between glacial and interglacial periods showed after entering the Quaternary, which facilitated the growth of plants in north and south as well as on the mountains and at the feet of mountains, and the modern grassland had been taking into shape. It can be seen that the grassland of China took shape in the Late Miocene (7,000,000 years ago) and expanded from Pliocene to Pleistocene, but the formation of Qinghai-Tibet Plateau was the latest one. Thus the present landscape approximately formed in Late Pleistocene (20,000~100,000 years B.P.).

It is no doubt that the occurrence and evolution of fauna is associated with the evolution of flora of grassland. It is easy to gain vegetal food, but compared with animal food on grassland, its nutritive value is relatively poor. Animals must intake a large amount of vegetal food as supplement. The grassland herbivores have a series of adaptation to this situation by adjusting their shapes and behaviors. Camels, deer, cattle, sheep and so on had developed ruminant food habits, which made them shorten feeding time and decrease the chance of being hunted by predators. Later, the food was digested bysymbiotic bacteria or anhistozoa within the rumen in appropriate time. Many animals of grassland run quickly to keep safe. We always see some groups that are expert in jumping and running, such as jerboa, rabbit, leaped rabbit, antelope, Mongolian gazelle, and deer, etc. Their light bodies and long legs can leave the predators far behind on the spacious grassland. Some minitype mammals (rodent) lived in underground cave to adapt to climatic seasonal variation of grassland and to evade predators; excavating caves will mix and loosen the soil, which plays a great role in the formation and succession of grassland communities. The eyesight of grassland animals is often acute. When observing ambient conditions with obstacles of bushwood, some minitype mammals often adapted to use hind legs to sit up to observe ambient conditions. There are various carnivorous animals on grasslands including large-scale lion, leopard, wolf, reynard, badger and eagle, etc. They contributed to maintaining the stability and balance of grassland's ecosystem.

Apparently, the development of grassland's flora has promoted the evolution of grassland fauna. But in return, the grassland fauna has also contributed to the formation of some plants. For instance, plants with thorn plants on grassland, beyond doubt, are the result that some large hoofed animals took branches and leaves as food. In the early twentieth century, some ecologists pointed out that "it is no doubt that grassland vegetation is formed under the influence of animals and it is capable of keeping its stability under persistent action of animals". At first, wild herbivore mightily affected grassland, which prevented grassland communities from degeneration because of excessive accumulation of the round layer. Afterwards, the livestock gradually replaced the large-scale wild animals. They have played important roles in the stability of grassland.

Ecological Functions of Grassland

The grassland ecosystem occupies distinctive geographic location on land, which covers extensive space on earth that is suitable for neither planting forests nor crop farming. The grassland is a green barrier to sustain ecological balance and important support system for creatures on the semi-arid area. It has established the natural foundation for sustainable development of human society.

Grassland is an important component of the life-support system on the earth and is closely bound up with human's existence and development.

As a life-supporting system, the basis of grassland is the biodiversity of populations and the complicated various network relations. The direct resource value of grassland is well known, but people failed to take more attention to its ecological functions.

Except for its grazing value, the low and sparse green covering on grassland is still a thin layer of "biological protective film" of the land surface. This biomembrane is composed of green plants and native algae, lichen, fungus that grow close to the ground, combined with herbaceous layer, they have many peculiar functions that guard against wind erosion, guarantee natural environment, adjust the ground temperature and protect biosphere. It has important significance to maintaining ecological environment balance of semi-arid climate and is irreplaceable by others. Many natural disasters occurred on grassland, such as grassland degradation, loss of soil and water, desertification and diseases and insect pests, were directly related to the damaged ecosystem structure and malfunction of grassland.

Study and practice have proved that the resource value and ecological functions of grassland are restricted by the structure of grassland ecosystem, self-renewal of biological component and self-reproduction. Therefore, exploitation and utilization of grassland should not affect its structure and function. The management of pasture resource should put emphasis on space-time balance of the number of livestock grass, and keeping the rational quota to make the grazing intensity maintainswithin the herding elasticity limitandto make the grassland obtains economic and optimum ecological benefits. Meanwhile, the comprehensive exploitation of other various natural resources of grassland must be based on protecting the ecosystem structure and integrating functions of grassland as well as maintaining the species biodiversity and stability of the system stability. This is the test standard of whether the grassland is rationally utilized and managed.

Glance of Grasslands of the World

The total area of the world's grassland is 24 million square kilometer, which accounts for 1/6 of the total land area. According to the characteristics of biological composition and geographic conditions, grasslands of the world can be divided into temperate grassland and tropical grassland.

Temperate grassland is distributed in mid-latitude zone of northern and southern hemispheres, such as Eurasian grassland (steppe), North American grassland (Prairie), and South American grassland (Pampas), etc. It is warm in summer and cold in winter. Spring season and late summer have obvious arid period. Because of low temperature and less rainfall, the grass group is very short and the height of most overground parts does not exceed 1m. The dominant species is cold–resistance xerophytic grasses. Calcification or swarding process isprevailing.

Tropical grassland is distributed in tropical and subtropical areas. In the middle of tall grasses (often up to 2~3 m height), it always disperse some low and short arbors which are called sparse tree grassland, also known as Savanna. It is warm all the year round, and annual rainfall reaches more than 1,000 mm. Under the influence of high temperature and less rainfall, the soil is barren. The two arid periods with in a year, and frequent occurrence of wild fire resticts forest growth.

Making a general observation of world grassland, we can come to a conclusion that the determinant factors which affect the distribution of grasslands are the combination of moisture content and heat. In geographical distribution, grasslands are located between moist forests and arid desert regions. The grasslands close to forests are in the sub-humid climatic region with lush grass and abundant species. Besides, island forests and bushwood appear, such as North American tall-grass prairie, South American Pampas, Eurasian meadow grassland andAfrican Savanna. The grasslands near deserts have short and sparse grass and simple species, which are mixed with undershrub or succulent plants, such as North American short grass prairie, and Eurasian desert steppes as well as extensively distributed typical steppes.

Eurasian grassland

The Eurasian grassland, also called the steppe, is the vast steppe ecoregion of Europe and Asia in the temperate grasslands where it is flat and open with good drainage and no water overflowing phenomenon in spring. It is the world's greatest and well-preserved grassland area. It starts from European Danube, stretching eastward like a strap, running through Hungary, Romania, Russia, Kazakhstan, Mongolia to Chinese Northeast Plain, and turns to southwest via Inner Mongolian Plateau, Loess Plateau and the southern margin of Qinghai-Tibet Plateau, which are continuous more than 8,000 km from east to west. The representative type of Eurasian grassland is various tussock-grass steppe. The Eurasian grassland can be divided into three subregions, namely the Black Sea-Kazakhstan grassland subregion, Asian central grassland subregion and Qinghai-Tibet Plateau grassland subregion. The Black Sea-Kazakhstan grassland subregion is located in west part of the Eurasian grassland. Its eastern boundary is located in the border of Xinjiang and Kazakhstan. Affected by Mediterranean climate, the climate is warm and humid in spring and hot in summer. There are two growth peaks

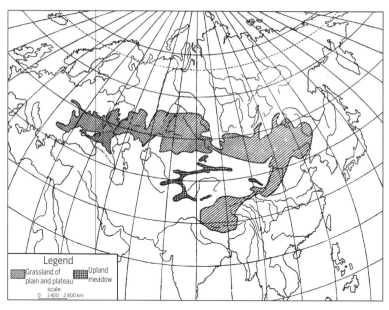

Sketch of Distribution of Eurasian Grassland (quoted from Vegetation of China)

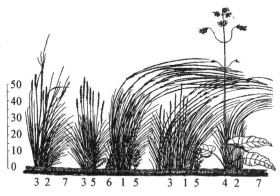

Vertical Projection of *Stipa* Steppe with Forbrich in Eastern Europe (quoted from Grassland of Soviet in 1940)

1-6 plant name 1. *Stipa lessingiana* 2. *S. capillata* 3. *Koeleria gracilis*
4. *Salvia nutans* 5. *Artemisia austriaca* 6. *Astragalus pallescens*
7. Ground layer

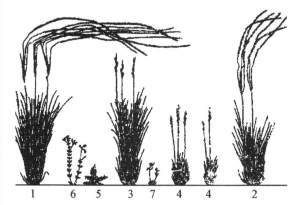

Vertical Projection of Grass Steppe with Stipa and Festuca as Dominant Species in Eastern Europe (quoted from Grassland of Soviet in 1940)

1-7 plant name 1. *Stipa ucrainica* 2. *S. lessingiana* 3. *S. capillata*
4. *Festuca sulcata* 5. *Pyrethrum millefoliatum* 6. *Cerastium ucrainicum*
7. *Erophila verna*

in a year. Ephemeral and ephemeroid plants grow very well in spring. The Asian central grassland subregion is located in the northeastern Eurasian grassland, including Mongolian Plateau Highland, Songliao Plain and Loess Plateau. Affected by the Pacific Ocean monsoon climate, winter is chill and summer is torrid. The growth curve of plants is unimodal and plant grows very well in summer. But it is lack of ephemeral and ephemeroid plants in spring. The Qinghai-Tibet Plateau grassland subregion is the world's highest grassland above the sea level. It is alpine steppe. The distribution of precipitation also is monsoon type. The temperature is very low all the year round. It has plant types that are adapted to alpine climate and alpine cushionplants in community often appear.

In the Eurasian grassland region, zonal differentiation is very evident caused by different situation of water and hot combination. From the area near forest to the desert region, the precipitation decreases gradually, but the heat increases, therefore the natural differentiation of forest steppe, typical and desert steppe appears.

The flora and fauna elements offorest steppe are complicated. The grass is luxuriant, and birch, poplar and oakery appear. The constructive species in the west is*Stipa joannis, Stipa lessingiana*, in eastern of Mongolian Plateau and the Northeast of China is *Stipa baicalensis, Leymus chinensis* and *Filifolium sibiricum* and in Loess Plateau is *Bothriochloa ischaemum*. The large-scale herbivores include moose, deer, roe deer and antelope, etc.

The typical steppes have typical semi-arid continent climate that are closer to inland especially than forest. The plant structure is simple that is dominated by xerophytic tussock grass and *Stipa*. In western part, the constructive species is *Stipa pennatae*, such as *Stipa lessingiana*, *Kazakhstan Stipa capillata*, *Stipa purpurea*, etc. The constructive species in the east is Sect. Leiostipa Dum, such as *Stipa grandi*s, *Stipa krylovii* and *Stipa bungeana*, etc. The familiar herbivores on the typical steppes are Mongolian gazelle, high nose antelope, ground squirrel, mouse rabbit and marmot, etc.

The desert steppeis the transition zone from grassland to desert. It is the aridest area of the steppe. The grass populations are lower and sparse with the height of only about 20cm. The constructive species are *Stipasareptan*, *Stipa gobica*, *Stipa breviflora* and *Ajania achilleoides*, etc. The main herbivores include sand mouse, owes an ear hamster, high nose antelope, etc.

North American Grassland

North American grassland (also known as Prairie), also called North American prairie, is the biggest grassland of western hemisphere that located in the center of the North American continent. It starts from southern Canada in the north, passes through the central of USA to the Gulf of Mexico; it starts from the eastern foot of Rockey Mountains and east to the west bank of US Great Lakes. It passes through from north to south alongmountainous region. It extends about 3,700 km. The east and west are long and narrow with the width about 700~800 km.

The rainfall of the Prairie gradually reduces from

east to west. Air temperature increases gradually from north to south. Thermophilic plants are in a dominant position in the south and cold resistant plants distribute in the north. Three types of grasslands including tall grass prairie, short grass prairie and desert steppe are distributed from east to west.

The tall grass prairie, also known as top grass area, covers a large area from Canada to the Gulf of Mexico which is about 1,000 km in length and 240~300 km in width. Its characteristic plants are mainly *Andropogon scoparlus* and *Andropogon gerarai*. But *Agropyron smithi* and *Stipa capillata* are widely distributed in arid areas. The tall grass prairie had become high-yield corn belt after being reclaimed for a long time. The short grass prairie is also called short grass country. It starts from the foot of Rockey Mountains and connects with tall grass prairie at west longitude 100°. It stretches from Canada in the north and to New Mexico State in the south, transiting to desert-grass zone. The natural vegetation gives first place to *Bouteloua gracilis* and *Buchloe dactyloides*.

The desert-grass area scatters in the south side of short grass prairie, and extends to the northern Mexico from the southwest of USA. Its latitude is below 1,300 m, the main grass is black grama and the main bushes are *Prosopic juliflora*, *Acacia farnesiana*, and

Deer on Short Grass Prairie of North America (Photographed by J.M. Suttie)

Virginiana small needle tree, etc.

The herbivores such as ancient elephants, mastodons, ancient camels, wild oxes and mooses as well as the cat predators such as pumas and cougars that taking herbivores as their food once inhabited on the North American grassland. The most famous one is American bison, which is also the largest hoof animal. Pronghorn is a unique hoof animal in North America and also runs fastest in the western hemisphere. The most famous rodent species is marmot or prairie dog. Birds on North American grassland include hazel grouse and *Sturnella neglecta*.

South American Grassland

South American grassland is also called pampas, which comes from Quechua language, meaning vast plain that gives first place to herbage. Now it refers to the extensive area stretching from Atlantic coast to Andes, which mainly lies in Middle East plain of Argentina at 32°~38° south latitude. The area is about 500,000 km^2, accounting for about 23% of the total area of Argentina. Its characteristic landscape is the small arbors shaping like an island that scatters on the vast grassland. Thereby, it is also classified into savanna. The topography of pampas is flat with abundant precipitation. The precipitation in the north amounts to 1,000~1,250 mm, the highest rainfall among all temperate grasslands. The rainfall decreases gradually southwest and to about 500mm at the southwest edge of the grassland. The air temperature is higher with intensive evaporation and dry climate. When the rainfall further reduces or temperate further rises, the grassland would be replaced by bushwood or savanna. The island-like forest which is composed of thorn *Celtis tetrandra* appears on light soil of the north. The dominant grassland community is composed of grass of *Bothriochloa ischaemum* and mixed with quite a number of forbs. At present, only very few primitive steppe community survives. Most of them have already been replaced by the farmland. Many large-scale bunch grasses dominated by South American *Stipa capillata* grow in the west and southwest. The animal husbandry of South American grassland only has a history of 400 years, but the productivity is consistently decreasing because of excess reclamation and heavy grazing in nearly a century. Vegetation composition was replaced by introduced species. Pampas has deep soil layer and natural fine grazing land which provides cattle farms for traditional Indian. The famous animals include the America ostrich, and *Tolypeutes tricinctus*, etc.

South African Grassland

South African grassland, also named Veld, refers to various kinds of open areas in South Africa. Some places are repeated from African savanna. It distributes mainly in the east and south of South African Plateau stretching from the south of Limpopo River to the southern coast. Due to higher elevation on highland, the comparatively mild temperate zone grassland climate with much rain in summer has been formed, which has been the old-land with the earliest record of human activities. The common characteristic of the

Elephants on African Savannah (Photographed by Xu Zhu)

Veld are less precipitation and higher temperature. Except a few areas, most areas suffer thin and barren soil. According to the altitude, the Veld is generally divided into high, medium and low position: the high-position veld with an altitude over 1,200~1,800m distributes in South Africa, Botswana, Lesotho, Zimbabwe and Zambia, where the endemic plant is *Bothriochloa pertus*; the Veld with an altitude over 600~1,200m distributes in the Cape of Good Hope and Namibia, where the dominant plants are heat-resistant plants, lofty perennial grass and forbs; and the Veld with an altitude of 150~600 m mainly distributes in Ruisiwaer, Swaziland and the southeast of Zambia. *Opopanax* and *Acacia cornigera* group alternative distribute with *Bothriochloa pertusa* in higher ground, while the *Bothriochloa pertusa* was replaced by hornwort herb, Euphorbiaceae plants and other succulents. Veld's animals lived in the veld including lion, leopard, elephant, giraffe, hippopotamus amphibius, oryx, sable and various birds, etc.

Savanna

Savanna is defined as: tropical or subtropical grassland containing scattered trees and drought-resistant undergrowth, which mainly distributes in Africa, South America and Oceania.

1. African Savanna

The African Savanna is a tropical grassland in Africa between latitude 15° North and 25° South along the north and south flanks of equatorial rain forest, which formed the belt extending from east to west. The belt width in the north is 400~550 km, and the length from east to west is 5,000 km. The belt width in the south is 200 km, and the length from east to west is 2,500 km. The total area is more than 800,000 km^2. The dominant plants are *Andropogoneae* and tall grass-

African Savannah (Photographed by Xu Zhu)

Tropical Savanna of South Africa (Photographed by Xu Zhu)

es of *Paniceae*. The scattered trees give priority to spiny *Acacia*. The tall scattered tree grassland dominated by *Pennisetum* and *Imperata* distributes in the area close to the tropical rain forest. Moving southward or northward with arid degree increasing, the grasses grow shorter, and the dominant plant is *Themeda hookeri*.

African savanna has many kinds of animals. Many of them are peculiar in Africa, such as African elephant, giraffe, zebra, wild ox, spotted hyena, African tyke, etc. Besides, there are many minitype mammals, birds, and a large amount of insects and other invertebrates.

2. South American Savanna

South American savanna distributes in the two sides between latitude 15° North and 25° South of the rain forest. A large area of savanna mainly distributes in Andes of and Orinoco of Venezuela, and Central Brazil. The others are located in the Colombian Orinoco plain, Amazon lowland, Bolivia, Pacific Ocean seacoast of Costa Rican and Mexico Gulf lowland. Compared with African savanna, the South American savanna is characterized by less and short arbors, besides, some grasslands even have no trees, such as the Venezuelan Llano. The dominant plants are *Panicum* grass and lofty cactus with thorns. It is difficult to find the native vegetation as it had been exploited.

3. Oceanian Savanna

The Oceanian savanna distributes in the north and west of the continent and it covers quite a large area. It still has *Acacia cornigera*, but has not trees and bushwoods with thorns. The dominant scattered tree is eucalypt. Most of herbaceous plants are tussock grasses which grow luxuriant in rainy season that can grow to 50~150cm at height. But they withered in dry season and the nutrition value is very low. The endemic animals are kangaroo and tree bear.

Chapter 2
Grassland Environment

Heaven
Lyrics and music composed by Tengger

The blue sky.
Clear lake water.
Green grassland.
This is my home.
The galloping horses.
Pure white flock of sheep.
And my lady.
This is my home.
I love you, my home.
My home, my heaven.
I love you, my home.

As a special type of ecosystem, grassland has its own especial environment. Make a comprehensive view of grasslands of the world, they are well distributed from temperate zone to tropical zone, and all inhabit the settled ecological location, which is between the wet forest and arid desert region.

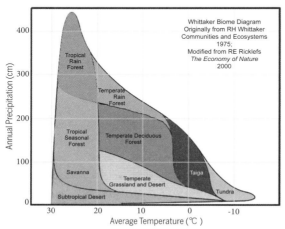

Sketch of Grassland Ecological location (quoted from Grassland of World)

The side close to forest boasts good moisture conditions, luxuriant grassesas well as island-forest or shrub; the area on the side of desert has less rainfall, and belongs to semi-arid and arid climate. It is covered with short and sparse grasses, simple component of classes, and mixed with some xeric arid undershrub and subshrub frequently; and the above-mentioned are semi-arid area typical steppe.

It should be noted that the combination of water and heat is the determinant factor of grassland distribution. Low temperature and less rainfall, and high temperature and more rainfall have the similar biological efficacy. The annual average temperature in the distribution area of savanna is 18~26℃, and annual rainfall is 1,000~1,500mm; but in temperate grassland, the annual average temperature is only -5℃ to 10℃, and annual rainfall is 100~600mm. There are significant differences, but both have the climate characteristics from semi-arid to semihumid, and have one or two arid periods in a year, which confines the growth of forest.

China's grassland is located in the north of temperate zone, which is the most important part of the Eurasian grassland. Its latitude stretches from approximately N 35° to N 52°, and longitude extends from E 83° to E 127°. In this vast area, the diversified ecological condition, together with the large difference of altitude (from 100m to 5,000m), making the grassland environment more complicated.

Physiognomy

Except the mountainous grassland in desert region, the most parts of our grassland distribute in Inner Mongolia Plateau, Loess Plateau and Qinghai-Tibet plateau, and only the northern parts extend to the Songliao Plain of Northeast China. The east of grasslands are surrounded by a series of mountainous regions, which are Xiaoxing'an Mountain→Zhangguangcai Mountain→Jibei Mountain→Lvliang Mountain→Minshan Mountain→east side of Qinghai-Tibet Plateau→Himalayas from northeast to southwest in proper order, and these mountains are the boundary of grasslandand forest regions of our country. The west side of steppes is surrounded by Helan Mountain, Qilian Mountain and Kunlun Mountain, and not far from several basins of the desert region.

Inner Mongolian Plateau (Photographed by Xu Zhu)

In the north of China's grasslands, Daxinganling Mountain extends from south to north, separating the Songliao Plain from Inner Mongolia Plateau; the Yinshan Mountains of eastwest trend rise in the central section, making the Inner Mongolia Plateau separating from Erdos Plateau and Loess Plateau. In a word, grasslands in China extend in strip shape from northeast to southwest, and move upward in ladder shape from northeast to southwest.

The northeast side of grasslands is Songliao Plain, which is embraced by these mountains: northern mountain area in Hebei in the southwest side, Daxinganling Mountain in the west, Xiaoxing'an Mountain in the north, and Zhangguangcai Mountain and Changbai Mountain in the east. The surrounding areas lean to the middle part.

The northwest of grasslands is Hulunbuir Plateau, which is a part of Inner Mongolia Plateau. With an altitude of 700~900m, it consists of the hills in the west of Daxinganling Mountain and the plateau itself. The ground presents undulate. The inland rivers include Hailar River, Wuerxun River, Kelunlu River and the Erguna River and its tributaries run through the plateau. Two inland lakes including Hulun Lake and Buir Lake distribute in the west. Hulunbuir Plateau is distributed with water system and partial areas have sand dunes.

Northeast China Plain www.eku.cc

The area from the south of Hulunbuir Plateau to the northern foot of Yinshan Mountains is the central part of Inner Mongolia Plateau, which is flat and vast-

Loess Plateau (Photographed by Wang Shunli)

with an average altitude of 1,000~1,300m and small Tenggeli desert located in the centre. Although there are few rivers, some billabongs and lake basins distribute in the plateau.

The south of Yinshan Mountains is Erdos Plateau, which occupies the southernmost of Inner Mongolia Plateau. It is an old land surrounded by the Yellow River on three sides. It has a peak elevation of 1,100~1,500m, the aeolian landform is widely distributed.

The south of Erdos Plateau locates Loess Plateau, the terrain of which is dominated by loess hills with an altitude of 1,500~2,000m. This area suffers from severe water erosion and tattered landform. The cutting depth of valleys reaches over a hundred meters. It is the area where water loss and soil erosion are most serious in China. The west of Loess Plateau lies in the Huangshui River valley, which has an elevation of above 3,000m, passes through the Riyue Mountain of Qinghai province and goes into Qinghai-Tibet Plateau.

Qinghai-Tibet Plateau is the highest and biggest plateau in the world which is known as the "Roof of the World". The grassland area occupies the central region of Qinghai-Tibet Plateau, its south is Himalayas, and north part is Kunlun Mountains, including most parts of Qiangtang Plateau, Yangtze River source region of southwestern Qinghai Province and southern Tibet. Gangdise Mountain, Nyenchen Tanglha Mountains, Alongganlei Mountain and Tanggula Mountains distribute in the area from south to north.

Qinghai-Tibet Plateau (Photographed by Xu Zhu)

Except the main part of the above-mentioned grassland, several mountainous grassland is distributed in Xinjiang, including Altai Mountain, Tarbahatai-Shawuer Mountain and Mount Tianshan. Altai Mountainruns northwest to southeast, and is the boundary mountain between China and Russia. The main peak height of Kuitun is 4,356m. The grasslandzone extends from 800m of mountain front to 1,200m (shady slope) and 2,100m (sunny slope). Tarbahatai and Shawuer Mountains are mountainous regions in western Junggar Basin with 4~5 peneplains. The highest altitude of peneplain is about 2,000m.The topography of mountainous region is flat. Tianshan Mountain is the watershed of Zhungeer basin and Talimu basin. It is steep with an elevation of more than 4,000m, and has the obvious mountainous vertical zone.The grassland zone occupies 1,100~1,700m (north slope) and 1,800~3,000m of the mountainous middle region.

Climate

China's grassland area is located in the mid-latitude inland of northern hemisphere mainly with temperate semi-arid climate, some areas belong to subhumid climate or arid climate, presenting obvious continent characteristics. It is dry and cold in winter under the influence of Mongolian high pressure air mass; and warm and rainy in summer influenced by oceanic monsoon climate. The area from east of Riyue Mountain of Qinghai to the northeastern Songliao Plainhas typical temperate climate characteristics;and the west of Riyue Mountain shows alpine climate characteristics, thereby it is divided into temperate grassland and alpine steppe. The upland meadow of desert region lies in the northwest China, where annual rainfall distribution is balanced under the impact of west wind all year round. The ecological physiognomy of China's grasslands is deeply characterized by the above climate characteristics.

Climate characteristics of temperate grassland

Temperate grassland is the main part of the northern pastoral area of China. It belongs to Pacific monsoon climate, where it is cold with less snow in winter and hot and rainy in summer.The east side of temperate grassland area borders forest area, and west side is neighbored with desert region. From east to west, the rainfall becomes lower but heat becomes higher, resulting in less and less available water. Degree of soil leaching has been gradually weakening, tepetate tends to close to the surface, the height, coverage, and abundance of grass become lower, and the composition of flora and fauna has changed, thus forest steppe (or meadow steppe), typical steppe and desert steppe occurred.

Besides the difference in east-west direction regions caused by different moisture, the difference also exists in south-north direction regions due to heat differences. Inner Mongolian Plateau is basically located in middle-temperate zone, and Loess plateau lies in warm-temperate zone, thus many obvious differences in terms of constructive species composition and development rhythm exist.

Climate characteristics of alpine steppe

The rainfall distribution of Qinghai-Tibet Plateau belongs to southeast monsoon type, but the temperature is very low all year round. It has no real summer, animals and plants grow under very low temperatures. The plants are very small, and cushion plants appear in the tussock. The constructive species is *Stipa purpure* of *Sect. Barbatae*, *Carex* moorcroftii and *Artemisia minor*.

The climate characteristics of alpine steppe are: abundant sunshine, strong solar radiation, low temperature, big daily range, small annual range, and distinct dry and wet seasons. The average temperature of warmest month is about 10℃, frost usually occurs in July or August, snow may even occur; and the average temperature of coldest month in winter is -5℃, and annual range of temperature is only 15~20℃. It is the region with the lowest annual range of temperature. But the day-night temperature difference is so great, usually can reach above 14℃, or even 20℃, much higher than that of the same latitude in mainland. The aforementioned is temperature situation plateaus, but in the south of plateaus, such as Yaluzangbu River, Kongque River and other valleys with the altitude below 4,000m, the annual average temperature is 1~8℃, accumulated temperature greater than or equal to 10℃ is 1,000~2,000℃, no-frost period is 70~150 days, and belong to temperate grassland climate. In fact, the steppe here is different from the alpine steppe of plateau, but similar to temperate grassland.

The annual rainfall of alpine steppe is 150~350mm, fasten on May-September, and the rainfall between May and September usually accounts for 90 % of the annual rainfall. It has much night rain with the night rain rate of 60%~70%. In every rainy season, it is nice in the daytime, but has thunderstorms at night, which is related to undulating terrain of Qinghai-Tibet Plateau. The temperature is high in the daytime, and water drops in the air are easy to vaporize, so it is cloudless; the temperature of ground decreases at night, cold air goes down along hillside, making the warm air of valleys rising up, and interchange of cold and warm air provides the condition for cloud and rainfall. In the growing season of Qinghai-Tibet Plateau, rainfall takes place at night, and sunlight is sufficient at daytime, plus the big difference of daytime temperature, it is good for photosynthesis of plants and accumulation of dry matter. Therefore, although the alpine steppe is short, nutrition value of grass is higher with a good reputation of "three higher and one lower" (namely, highercontent of protein, fat and vitamin, but lower fibre).

Climate characteristics of upland meadow in desert region

Several huge mountains and foot of the mountain in Xinjiang belong to desert climate, and the grassland-climate is the product of mountainous vertical zone. As mentioned above, this area is influenced by west wind and moist flow of the Arctic Ocean, the distribution of annual rainfall is balanced. Take Tacheng as an example, its annual rainfall is 280.3mm, of which, rainfall in spring is 69.6mm; rainfall in summer is 84.2mm; 81.5mm in autumn; and 45.0mm in winter. The rainfall in spring is obviously more than other grassland areas.

At the foot of the mountain and flat areas, annual average temperature is mostly above 5℃, but annual average rainfall is below 150mm. The desert vegetation mainly consists of shrubs and sub-shrub grow here. Temperature begins to decrease above the baseband. Within the scope of mountainous grasslandzone, annual average temperature is 0~5℃, annual rainfall is 150~500mm, annual active accumulated temperature which may begreateror equal to 10℃ is 2,000~3,000℃, having the temperate and semi-arid climate characteristics. Going up from steppe zone, temperature declines sharply, and then it enters into subalpine and tierra.

Soil

The development of grassland soil is restricted by climate, topography, hydrology and biological activities, which can be divided into two series, including zonal (obvious field) soil and azonal (dormant field) soil. The soil of zonal grassland in moderate temperate zones can be divided into several kinds, including black soil, chernozem, chestnut soil and brown calcic soil; the corresponding soil type of warm temperate zone includes Dark loessial soil and sierozem; upland meadow includes mountainous chernozem, mountainous chestnut soil and mountainous brown calcic soil; and soil types in alpine steppe include alpine meadow soil and alpine steppe soil, their common characteristics are:

(1) Under the impact of semi-arid to subhumid climate of temperate zone or alpine, the process of soil-grow-grass is the dominant process, the obvious humus horizon has been formed under the surface soil with high organic content, usually can reach 1%~5%, and up to 10%~15%, becoming one of the most important storerooms of carbon material.

(2) The eluviation is weak, some soluble salts such as calcium carbonate deposits under a certain depth of soil, and forms the tepetate.

(3) In initial conditions with less man-made interference, the surface has more or less continuous litter material, but under the influence of grazing or other human activities, the litter material could completely disappear.

Grassland soil of moderate temperate zone

1. Black soil

Distributed in the northeast of moderate temperate grassland, the black soil is mainly found in the hills between the Greater and Lesser Khingan Mountains and Songnen Plain. It is subhumid climate with very abundant vegetation specie, andone of main kinds is weedy meadow (locally known as tessellated meadow).

Because of the high rainfall and strong eluviation, black soil is the only type that has no carbonate sediment and presents weakly acid (pH 5.6~6.6) at the section. Humus layer at surface soil is thick and black, the content of organic is 5%~10%, and up to 17%, and is the most fertile soil type in nature.

2. Chernozem

Chernozem is the most representative soil type which belongs to forest steppe zone of temperate grassland. The vegetation is thick meadow steppe. Soil humus is so deep to a depth of 60~90cm. Its content of organic is 3.5%~7%, and the highest can reach 10%~12%. It is grey black or dark grey, and has obvious granular structure. The tepetate is deeply buried at a depth of 60~90cm, the content of calcium carbonate is also low. The soil is neutral (pH 6.5~8.0), and become higher with increased depth of soil. Chernozem is very fertile soil type in nature, some lands have been converted to farmland, and have become grain production base dominated by wheat.

3. Chestnut soil

This is the most typical steppe soil, consistent with the typical steppe zone, forming chestnut soil strip. It is located in temperate semi-arid climate zone. Caespitoso-Graminosa is the main vegetation, and the thick degree decreases. Humus layer is maroon or taupe brown with the thickness of 25~45cm, which is thinner than chernozem. The content of organic is 1.5%~4.5%, or even 5.5%. The tepetate is located deeper than chernozem, which generally appears at a depth of 35~50cm soil, and its thickness is 20~60cm. The content of calcium carbonate in tepetate is much higher than that of chernozem, which is 10%~30%. The soil presents weakly alkaline and alkaline (pH7~9), and increases with increasing depth. The fertility of chestnut soil is moderate, which is suitable for farming, but is usually restricted by moisture condition, not suitable for large-area reclamation.

4. Brown calcic soil

This is the most arid type of grassland soil, which is found in the neighboring areas of grassland and desert. It is in arid climate region. The vegetation is desert steppe, which is low and sparse. The earth's surface is short of ground layer, which under the wind action, generally occurs "gravel face". Humus accumulation presents weakness and its color is brown or pale yellow brown. The content of organic is 1.2%~1.8%. The location of tepetate is higher, generally 20~30cm depth; the thickness of tepetate is also thin, generally no more than 20~30cm. Full section presents alkaline, pH9~9.5. The bottom always appears a little gesso, which also explains that the eluviation is becoming weakening. Brown calcium soil is unfit for cultivation, and if there is no irrigate conditions, crops can't yield good harvest.

Grassland soil of warm-temperate zone

1. Dark loessial soil

Dark loessial soil is a soil type formed under the condition of warm temperate zone forest steppe, and mainly found in loess plateaus in north of Shaanxi province, and northwest of Shanxi province. Humus layer structure of dark loessial soil is loose and deep, which can reach more than 80cm in depth, presenting grey brown color. Because of the high temperature, its decomposition process is fast, content of organic is not so high, generally is 1%~2% in surface layer, much lower than that of chernozem. Carbonate presents fake-mycelial and distributes along the surface of gap. It is believed that dark loessial soil is the variant of chernozem under the condition of warm temperate zone. At present, most of dark loessial soil has been reclaimed.

2. Sierozem

Sierozem is the dominant soil type of warm temperate desert steppe, and has extended into typical steppe zone. Its section differentiation is not so obvious, and humus layer presents yellow brown and grey color. Its constructive property is bad and the content of organic is low, generally only 0.5%~0.9%, and not exceeding 1% to the highest. But as humus goes deep, transiting adown is not so obvious. Tepetate is located at 50~70cm in depth. Full section shows alkaline (pH is above 9, a slight increase from top to bottom). The bottom of soil has a little sodium carbonate.

Soil of alpine steppe

1. Alpine meadow soil

Alpine meadow soil is a special soil type which is formed in Qinghai-Tibet Plateau under cold and semihumid condition. Its vegetation is alpine meadow and meadow steppe. Basically, soil-forming process is grassing process. The root system of grass entwists on the surface layer of soil, forming tighten felted greensward. Each layer underground has a big proportion of gravel. The calcium carbonate accumulation has been formed in the section with different degrees. The full section shows alkaline (pH 6~7.5).

2. Alpine steppe soil

Alpine steppe soil is asoil type formed under the cold and semi-arid condition and is connected with alpine steppe. Grassing process of soil is weaker than

Xilin River

that of alpine meadow steppe. Accumulation of organic residues and humus become less, so it is difficult to form obvious divot. The content of organic on the surface layer is 1.5%~3%. Calcification process is obvious. All layers have strong lime reaction (pH 7.7~8.7).

Hydrology

Compared with desert ecological area, the hydrological condition of grassland is that surface water and groundwater resources are abundant, but the space-time collocation of water resources presents obvious imbalance characteristics. The distribution of drainage is generally restricted by geological structure and close related to topography and rainfall condition. The external drainage of China's grassland, from north to south includes: Heilongjiang water system, Liaohe River water system, Luanhe River and Yellow River water system, etc. In general, river network of eastern, southeastern boondocks is developed, and endowed with abundant water resources. However, the spacious dry steppe and desert steppe zones are blind drainage in central Asia, where density of river network is small with a low flow rate. Most water systems become dry valleys, or inland salt lakes with different sizes due to impeded hydrological conditions. The change ranges greatly between wet season and dry season, which become the most important restraining factor for agricultural and animal husbandry production and development.

Khalkhyn River Originated from Arxan Mountain Flows into Buir Lake (Photographed by Yong Shipeng)

Chapter 2 Grassland Environment **27**

Namtso Lake in Tibet

Qinghai Lake Region

Qinghai Lake Bird Island

28 Grassland of China

Ergun River

Swamp of Qinghai-Tibet Plateau (Photographed by Xu Zhu)

Chapter 3
Grassland Vegetation

Father's Prairie, Mother's River
Written by Chinese poetess, Xi Murong.

Father used to describe the fragrance of the prairie.
A scent that followed him to the edges of the world.
Mother always spoke of the turbulence of the river.
Raging through the Mongolian Plateaus, my distant home.
Now that I finally come to see this great land.
Tears rain down my face as I stand on these fragrant prairies.
The river sings of the prayers of the forefathers.
Blessing the prodigal son to find his way home.
Ah, father's prairie.
Ah, mother's river.
Though I can no longer express them in my mother tongue.
Please accept my feelings of sorrow and joy.
I am also a son of the prairie.
There is a song in my heart.
It sings of my father's prairie and my mother's river.

Species Composition of Grassland Vegetation

According to incomplete statistics, China's temperate grassland regions have more than 3,600 species of seed plants, which belong to 125 families, of which, Inner Mongolian grassland has 1,519 types of seed plants, belonging to 94 families and 541 generas. The sequence of families which contains much more species are: *Compositae* > *Poaceae* > *Leguminosea* > *Ranunculaceae* > *Cyperaceae* > *Rosacea*, and the sequence of genera which contain much more species are: *Carex* > *Artemisia* > *Astragalus* > *Polygonum* > *Salix* > *Potentilla* > *Allium*. The category of grassland is significantly different with adjacent forest zone and desert region.

Temperate grassland vegetationwas gradually evolving and developing from the process of long steppification after Oligocene. Xerophile has strong adaptability to the variance of climate in angiosperm, which created conditions for the formation and evolution of grassland ecosystem. Especially the polarization and development of *Stipa*as well asthe figuration and reinforcement of its constructive effect, contribute to the development of grassland ecosystem to flourishing period. Therefore, it was appropriate that people called temperate grassland in Eurasia as the "kingdom" of *Stipa* steppe. *Stipa* can be found all over the continent grassland region, about 300 species in the world, 27 species in China, in which 16 species are constructive

Stipa sareptana-Constructive Species of Desert Steppe in Kazakstan(quoted from Eurasian Grassland in Russian)

species of grassland community, including 8 sections.

In the composition of grassland vegetation, *Stipa baicalensis*,*Stipa grandis*, *Stipa krylovii*, *Stipa bungeana* and *Stipa capillata* are basic constructive species, which make up of the formation of temperate grassland. Sect. *Leostipa* is a relatively old species in *Stipa*, formed in mountainous region in late tertiary, and gradually expands to plain as the climate changes.

In addition toSect. *Leostipa*, Sect.S*mirnovia* and Sect, B*arbatae* also play very important roles in

Festuca valesiaca −A Characteristic Species Widely Distributed in Zonal Steppe in Kazakstan(Quoted from Eurasian Grassland in Russian)

grassland vegetation. *Stipa klemenzii*, belongs to sect. *Smirnovia*, stems from some islands and islets on the east bank of ancient Mediterranean, is a specialized group on modality and ecology, in which *Stipa gobica*, *Stipa klemenzii*, *Stipa glareosa* and *Stipa caucasica* are constructive species of desert steppe, widely distributed in desert steppe and desert mountainous region. *Stipa breviflora*, *Stipa orientalis* and *Stipa purpurea* belong to Sect.*Barbatae*. *Stipa breviflora* and *Stipa orientalis*, are constructive species of desert steppe, widely distributed in the center of Asia. *Stipa purpurea* is the most important constructive species of alpine grassland in Qinghai-Tibet Plateau. Besides, *Stipa subsessiliflora* and *Stipa subsessiliflora var. basiplumosa* (belonging to sect. *Pseu-doptlagrosits*) are constructive species of alpine steppe.

In addition to *Stipa*, *Festuca* is the dominant composition of desert-mountain meadow and sandy grassland. There are about 100 species of *Festuca* around the globe, and 23 species can be found in China. In which, formation *Festuca ovina* constituted by *Festuca sulcata*, *Festuca ovina* and *Festuca alatavica* etc, is an excellent pasture.

Agropyron, *Koeleria*, *Cleistogenes* and *Poa* are caespitose, commonly as companion species of formation *Stipa* and formation *Festuca ovina*, and can translate into dominant species at sandy grassland and overgrazing section, forming sandy grassland.

Bothriochloa ischaemum(belonging to *Bothriochloa*)is a representative formation in warm temperate grassland region, it can spread to dry hillside of deciduous broad-leaved forest, and become the constructive species of secondary grassland.

Aneurolepidium chinense(belonging to *leymus*), is high-quality forage, can take the role of construc-

Formation Aneuvolepidium chinense (Photographed by Xu Zhu)

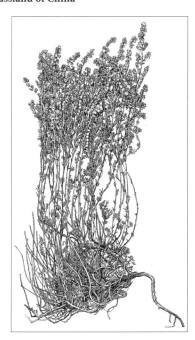

Artemisia frigid—North American Species Widely Distributed in Palaeotropical Zone(Quoted from Eurasian Grassland in Russian)

tive species and form a characteristic grassland type distributed in Northeast China Plain and Inner Mongolian Plateau, which is the most precious and natural mowing grassland. And *Aneurolepidium angustus* is the dominant species of wet salt habitat, but it only occupies a small area.

Forbs which make up grassland vegetation are very abundant, but there are limited constructive species, most of which are characteristic vegetations of specific formation or ecological index species of special habitats. *Composite* is the biggest family in grassland region, including 70 genera and 244 species, in which *Artemisia* takes a relative important role in the composition of grassland vegetation, contains rela- tively abundant species, about 50 species, and has evident ecological variationization, including xeric (such as *Artemisia frigida*), mesogenous (such as *Artemisia gmelinii*), sabulose and saxicolous species. *Artemisia* is the companion species in grassland community, but it can bring convergence effect under some outside force (such as continuous overgrazing), enlarge population viability, increase individual numbers, gradually occupy space, transform some grass steppe community into *Artemisia* community, and become a pattern of manifestation of desertification. *Artemisia frigida* and *Artemisia sacrorum*, which widely distributed in meadow steppe and dry steppe, belong to this phenomenon. Large area *Artemisia* community forms *Artemisia* steppe, which is paratactic with grass steppe.

Filifolium, only contains one species—*Filifolium sibiricum*, is a particular genus in mountainous region of East Asia. The forb grassland, which takes *Filifolium sibiricum* as constructive species, extensively distributed on the east and west sides of Daxinganling, extends northward to *Stipa baicalensis* steppe, becoming a particular formation in Davurica- Mongolia grassland.

Family *Leguminosea* contains 25 genera, 123 species, and is the third biggest family in grassland region. People often pay more attention to it because of its high nutrient value and rhizobium in the root. But in grassland community, *Leguminosea* plants often serve as companion species with no more individual numbers and low biomass. Only *Caragana*, a landscape type shrub, distributed on temperate grassland, scatters on grass steppe as background, forming a varied thicketization-grassland, and developing into sand shrubs on fixed sand. *Caragana* contains 16 species and are distributed in grassland region and adjacent region, which constitutes an integrated ecological evolvement combined sequence composed by mesogenous dungarunga and xeric, strong-xeric and cold-xeric dwarf shrub. Besides, *Astragalus*, *Oxytropis*, *Vicia*, *Hedysarum* and *Glycyrrhiza* etc. are also important germplasm resources.

Floristic Geographical Elements of Grassland Vegetation

Floristic geographical distribution of grassland vegetation is relatively complicated. Based on number of species, the basic geographical elements of Chinese temperate grassland flora are East Asia species, temperate Asian species, North Temperate Zone species, old continent temperate species, and Mongolian species. The diversity of the floristic geographical elements in grassland region reflects the basic temperate characteristics of Chinese grassland vegetation and its relation with adjacent floristic geographical elements on occurrence.

Specifically with regard to the grassland vegetations which have regional significance, most of the grassland plants are particular species in arid and semi-arid region in central Asia. There are plenty of common species which appear in grassland community, and the floristic geographic element is even more complicated. Due to space reasons, we will not enter into details here.

Life Form of Grassland Vegetation

Life form (adaptive convergence) refers to the phenomenon of similarity of different kinds of biology on configuration, physiology, development and adaptation fashion and approach by living in the same or similar environment for a long time through variation, selection and adaptation. The life form composition of grassland vegetation is different from other vegetation types. Perennial, long nutrient period and polycarpic xerophytic herbaceous plants are the main body of grassland vegetation, in which bunch grass takes the most important role in ecological construction, followed by rhizome grass and axis-rooted type forbs. Besides of perennial herbs, annual, biennial, semi-shrub, dwarf shrub and shrub are also included in life form. There is no forest landscape on grassland, arbor only are distributed in special habitats in grassland region, such as in mountains and dunes. The distribution of shrub is relatively universal, forming particular shrub-grassland landscape.

Tussock grass As the main body of grassland vegetation, tussock grass has special adaptation superiority to temperate arid and semi-arid continent climate and the trample and grazing of hoofed animals compared with other life forms. Sizeable slice of tussock base (about 5~9 cm) is embedded in the surface layer of soil, forming a firmly deadwood strip to wrap the renewal bud, which is not only propitious to the accumulation of rain, snow and dust particulate, but also can improve local water and temperature condition. Developed fibrous rootsystem can form a huge moisture and nutrition absorbing surface on humus layer, which can effectively use the limited moisture in the soil. Sclerotic stems and branches, curly lamina and particular dissection structure allow lesser water-holding capacity and intense transpiration to achieve powerful absorbing capacity and adaptive mechanism of enduring organization dehydration. The study of comparative plant ecology shows that the response of grassland vegetation to arid and high temperature of soil and atmosphere is achieved by the correlation between adaptation and high intensive physiological process, and the resistance of desert plants to bad living environment is achieved by decreasing metabolic process. In terms of photosynthesis, the prairie grass is remarkably adaptive to continent climate conditions. Observation in the field found that *Stipa gobica* could maintain relatively high photosynthesis and absorb CO_2 with the consistence of 22%~56% when the temperature arrives to 0~ -10℃. All of the facts show that the adaptation of grassland vegetation to low temperature and drought shows strong plasticity and diversity, which are physiological features not possessed by mesophytic herbage. Flowering habit and seed distribution of grassland vegetation are also distinctive. Most of the grassland vegetations are chasmogamous anemophilous plants, flowering and pollination often keep close connection with air temperature. But we discovered that the species of *Stipa* are in fact cleistogamous plant, all of the anthotaxy lies on the top of reproductive branches, wrapped by flag leaves closely before the maturation of the seed. After the seed maturation, the flag leaves dry out and unfold, and the caryopsis of *Stipa* is embedded in soil with the help of wind. *Cleistogenes* plants are steady cleistogamous plant, which often forms lateral branches in the axil of flag leaves. These are particular adaptation characteristics of grassland vegetation formed in the long evolutionary process.

Besides, grassland vegetation also formed special adaptabilities in terms of community structure, seasonal development and ecological functions. For example, tussock grass steppe does not form green omnidirectional coverage, by which we can distinguish it with typical mesogenous meadow steppe. The conflict of water and nutrition contest between plants could be eased strongly because of the existence of intervals between thickets.

Orientation observation shows that the seed

regeneration of grassland vegetation is very difficult, but the age composition of basic population still keeps normal state and belongs to unsteady and regenerative equative steady state. Age structure of constructive species (such as *Stipa*) is often characterized by dominance of sexually mature and middle age population, which is an efficient adaptation strategy to bad regenerative condition.

Like other temperate vegetation forms, the seasonal development rhythm of grassland vegetationcommunity is very evident. The most distinctive characteristic is the lengthy winter dormancy besides the growth and development stage. The grassland in Mediterranean climate region also has a period of semidormancy in summer. The duration of winter dormancy is varied from place to place, the growth period of moderate-temperate steppe distributed in the north of Yinshan Mountains is shorter compared with warm-temperate steppe distributing in Loess Plateau, the cold-season dormancy prolongs about 15~25 days.

Many years' observation of fixed position proves that the interannual changes of grassland community, including the quantitative relationship between population structure and production level, are restricted by 11 to 13-year periodic fluctuation disciplinary which is associated with amplitude of solar activity. It was pointed out that in grassland community that the maximum above-ground biomass and grass population foundation are offered by tussock grass. This type of plant can be divided into two ranks, the first rank is composed by tall tussock grass, some species of *Stipa* is in dominant position;the second rank is composed by dwarfish tussock grass and *Carex*, the representative genera are *Festuca*, *Cleistogenes*, *Agropyron*, *Koeleria*, *Poa*, *Carex* and some pint-sized species of *Stipa*, etc. This structural feature of grassland vegetation is a common characteristic of Eurasian grassland vegetation, but has some discrepancy in different regions. The Black Sea-Kazakstan grassland is dominated by large-scale closely-cluster grass. In middle Asia grassland, large-scale, medium-scale and pint-scale closely-cluster grass occupy the dominant position under different eco-geographic conditions, reflecting the diversity of grassland in ecological distribution.

Bunchgrass Bunchgrass often appears in meadow steppe and some forb-tussock grass steppe, and the representative species are *Phleum phleoides*, *Helictotrichon schellianum* and *Poa stepposa,* etc. *Cleistogenes squarrosa* and *C.songorica* are subdominant species of dry steppe and desert steppe, and can become dominant species under local conditions.

Rhizomatic grass Rhizomatic grass often has unconspicuous effect in the community of Eurasian grassland region, but it becomes the dominant synusia in Northeast Plain and eastern grassland in Inner Mongolian Plateau, constituting a type of particular rhizomatic grass steppe. Rhizomatic grass steppe is a very important grassland type, and its representative species is *Aneurolepidium chinense*.

Dicotyledon Dicotyledon plays a less important role in grassland compared with tussock grass according to its importantance value. Different families and genera of dicotyledon occupy a definite proportion in community, but the quantity and composition gradually decrease with the decrease of precipitation from meadow steppe to desert steppe. Forbs contain many different growth-form plants, including axes-rooted plants (such as *Dianthu, Centaurea, Salvia officinalis,* and *Astragalus*, etc), rhizomatic plants (such as *Veronica*) and tuber plants (such as *Trifolium*), etc.

Ephemeral plants, ephemeroids, and annual and biannual long nutritional plants   In terms of the general characteristics of grassland vegetation, we take perennial, long vegetative, polycarpic and tepidity xerophytic plants as particular characteristics of grassland vegetation when defining the types of grassland vegetation. But as a matter of fact, many ephemeral and short vegetative plants are included in the floristic composition of grassland vegetation.

Short vegetative plant refers to a kind of herbage with an ultrashort life cycle. They complete the whole life cycle in a very short time (1~2 month), of which, the plants completing the whole growth and development process in the limited time within a year is called ephemeral plant or ephemeral; and perennial short vegetative plant is called ephemeroids, which are considered as one of the most particular characteristics of grassland vegetation in Mediterranean climate region. The representative plants of this kind of herbage include *Tulipa, Gagea, poa bulbosa,* etc. Filed investigation shows that ephemeroids such as *Tulipa uniflora* and *Gagea pauciflora* are also widely distributed in Mongolian Plateau grassland region, but its coenological effect is much less obvious than in Mediterranean climate region, reflecting some historical similarity and actual difference in these two adjacent climate regions.

Annual and biannual long vegetative plant is the common companion species of various steppe communities, but the population shows an obvious trend of expansion in desert steppe and overgrazing degraded grassland. The common species includes *Artemisia scoparia, A. sievensiaria, Salsola collina, Eragrostis minor, Setaria viridis, Chloris virgata, Aristida adscensionis* and *Pappophorum borealis,* etc. These plants live through the winter relying on seed, grow and flower rapidly in rainy season and complete the growth cycle. The biomass fluctuates with the precipitation interannually. The plants flourish in the rainy season, forming the background, and failure to thrive in the rainless season, producing very little biomass, and becoming an unstabilizing factor of grassland community structure. This phenomenon tends to be more various as moving to dry region. Annual and biannual plants such as *Chenopodiaceae, Brassicaceae* and *Asteraceae* often act as pioneer plants in immobilizing quicksand.

Parasitic and semi-parasitic herbaceous plants   All grassland plants outlined above are autophytes. Besides, in some communities, there is a small quantity of parasitic and semiparasitic plants, which reflects the complicated nutritional relationships among plants during evolution. The common parasitic plants include *Orobanche pycnostachya, O.coerulescens, O.arnurelasis* (parasitize at the root of *Artemisia*) which belong to Orobanche cumana Wallr, *Cuscuta chinensis*, *Cuscuta japonica*, *Cuscuta europaea* (parasitize family *Leguminosea, Asteraceae* and *Chenopodiaceae*) and *Cynomorium songaricum* of *Cynomoriaceae* (parasitize at the root of family *Nitraria tangutorum* Bor.), etc. The quantity of semiparasite is lesser, including *Thesium chinens*, *Th.longifolium* and *Th.refractum* etc belonging to *Thesium (Santalaceae)*, commonly acts as companion species, and mainly distributed in forest-steppe region.

Subshrub and dwarf subshrub Subshrub and dwarf subshrub are another particular life form formed in arid and semi-arid continent climate under natural selection. The characteristics of this type of plant are as follows: when the above-ground plants of grassland wither and keep in semi-dormancy state in cold season due to low temperature, such kind of plant can live through the winter relying on its high cold resistance capacity and its high lignification in the section of above-ground tress approaching to $1/3 \sim 2/3$ of the ground, and the renewal bud and underground root system can keep living, and only non-lignification part at the top of the tress die back, therefore it is called subshrub (or dwarf subshrub) or semi-wood and semi-herbage plant. The representative species of subshrub and dwarf subshrub are some species of *Artemisia*, *Hippolytia* and *Ajania of Asteraceae*, *Kochia of Chenopodiaceae* and *Thymus of Lamiaceae*. They often act as companion species, but in extreme environments, they can develop into dominant species, forming peculiar subshrubgrassland vegetation relying on the peculiar stress resistance and adaptive capacity to the habits of drought, quicksand, semi-quicksand and superstrong erosion. The representative formations are *Artemisia frigida* steppe, *Thymus quinquecostatus* steppe and *Artemisia ordosica* Formation, etc. Commonly, subshrub synusia is known as dominant structural feature of desert steppe. The enhanced role of dwarf subshrub in grassland community is regarded as the symbol of desertification of grass steppe.

Shrub Shrub belongs to low-phanerophytes, and the representative species include *Spiraea hypericifolia of Spiraea chinensis*, genus *Amygdalus pedumculata of Amygdalus* and *Prunus sibirica of Prunus*, etc, in addition to *Caragana* listed above. These shrubs can develop into dominant species in upland meadow and dune, though they do not have so obvious landscape function in grassland zonal vegetation comparing with *Chrysolophus*. In particular, *Prunus sibirica* steppe, which is a characteristic species, is an important local grassland resource in the mountain-foot plain of east Daxinganling. Most shrubs have developed root system and asexual reproduction capacity. When the normal growth of herbage is restrained by overgrazing, the shrubs with the capacity of rapidly enlarging population, such as *Caragana microphylla*, can form steppe shrub on degraded grassland. This phenomenon is universal in Xilin Gol Grassland.

In nature, grassland belongs to non-forest landscape, but some special habitats such as mountain, dune and gully in grassland regions are the "refuges" for some arbors, where different kinds of fragmentary association have been reserved. *Juniperus rigida* woodland, which has obvious grassland characteristics, has been retained in Yin Shan Mountain.So far, and *Ulmus pumila* woodland (the main body are *Ulmus pumila* and *U.macrocalpa*) is often seen in dune. *Pinus tabulaeformis* and *Picea mongolica* often exist in Hunshandake Dune, which become the important symbol and scientific basis of studying and analyzing the climate change in north China.

Bryophyte, lichen, Fungus and algae According to the investigation, 92 species of *bryophyte* are distributed in typical and meadow steppes of Hulunbuir and Xilin Gol, but *bryoflora* is relatively insufficient in grassland community, only has some species, such as *Bryum*, *Weissia* and *Barbula*. Abundant *bryophyte* plants distribute in the forest community of mountain forest-grassland region, some species such as *Amblystegiaceae*, *Bryaceae* and *Hypnaceae* are the main body. 31 species *bryophyte* plants are discovered in the south of Yinshan Mountains and Erdos, including *Grimmia*, *Barbula*, *Didymodon gigateus*, *Reboulia*, *Plagiochasma*,*AloeSphagnum*, *Bryum* and *Tortula Hedw*, etc. But only 8 *bryophyte* species distribute in desert steppe, including *Barbula*, *Tortula Hedw*, *AloeSphagnum* and *Grimmia*.

Lichen is mainly found in dry steppe and desert steppe, and the common species are *Parmelia vegans, P. ryssolea* and varied parti-color crustaceous lichen.

There is a great variety of Funguswhich mainly appears in semi-arid and semi-humid climatic regions. Many edible Fungus, such as dried mushroom, straw mushroom and *Lepista Sordida* are distributed in *Leymus Chinensis* steppe of eastern Inner Mongolia.

Autumnal Scenery of Wild Apricot Shrub Steppe (Photographed by Yong Shipeng)

Ulmus pumila Woodland After Herbaceous Plants Destroyed (Photographed by Yong Shipeng)

Rock Surface Multicolor Crustaceous Lichen (Photographed by Yong Shipeng)

Stratonostoc commune and *Nenalonostoc flagellifore* of *Cyanobacteria* are widely distributed, and have obvious geographic substitution phenomenon. The former mainly appears in semi-arid typical steppe and meadow steppe, and the latter distributes in desert steppe and salinization land in half-desert region.

Ecological Group of Grassland Vegetation

Plant ecological group refers to the plant group of various ecological habits and different ecological types which formed through long-term adaptation to different environments. The analysis of ecological group is the foundation of understanding grassland characteristics and evaluating its ecological functions.

Plants which compose the grassland vegetation grow in different environments. They formed various adaptive characteristics to water condition, soil salinity and caloric condition in the process of natural evolvement. Ecologists divided plants into different ecological groups according to their adaptive capacity to ecological factors or living habits using the method of comparative ecology.

Water condition Water condition is the basic ecological condition which restricts the distribution of grassland vegetation. Plants are divided into mesophyte, mesoxerophytes, xerophyte, strong xerophyte and extreme xerophyte according to their adaptation to water factors. Mesophyte and mesoxerophyte are the characteristic species of meadow steppe, strong xerophyte and extreme xerophyte only exist in desert steppe, and mesophyte is the dominant species of typical steppe.

Thermal gradient The plants of grassland vegetation can be divided into thermophilic species (such as *Stipa bungeana* and *Bothriochloa ischaemum*), mesothermal species (such as *Stipa grandis* and *Stipa klemenzii*), cold resistant species (such as *Stipa baicalensis* and *Filifolium sibiricum* etc.) and alpine climate resistant species (such as *Stipa purpurea* and *Leucopoa albida*) according to the change of latitude and elevation.

Soil Soil is the foundation of plant existence, and it is also the source for plants to get access to water and nutrition. Calcic soil (chernozem, chestnut soil and brown calcic soil) is the main soil in grassland region, and many plants in this area are calcicole. The plants adapt to the salinity in habitat by different ways, and are divided into salt-tolerance plants and euhalophytes, the former (such as *Aneurolepidium chinensis*, *Chloris virgata* and *Iris ensata thumb*, etc.) adapt to saline environment through different physiological mechanism and the regulation of intracellular and extracellular osmotic potential. The euhalophytes are able to survive in high-concentration salt habitats through salt excretion, polysalt and salt rejection. The cell of salt excretion plant has very high permeability to dissoluble salt, can absorb ions, and the salt is not accumulated completely in the body, but the superfluous salt ions are excreted by secretory, such as *Limonium bicolor (Bunge) Kuntze* and *Polygonum sibiricum var. sibiricum*. Polysalt plant refers to the plant that needs high-concentration salt genetically to achieve normal growth and development. This type of plant absorbs a mass of salt ions to keep in body, in order to increase the osmotic pressure, and hypertonicity is propitious to maintain the water potential of the cell and the life activity so as to promote growing. *Suaeda corniculata*, *S.glauca*, *Salicornia europaea* and *Kalidium foliatum* commonly found in saline-alkali lakeside of grasslandare the representatives of salt accumulation plant. On contrary, the permeability of salt of salt-rejection plants is very low, and its permeation is adjusted by organic acid, amino acid and sugar in the cell. The dominant species widely distributed in salting wet land of grassland is represented by *Puccinellia tenuiflora*, *P. chinampoensis* and *Hordeum brevisubulatum*.

Psammophyte Psammophyte refers to the plant which grows in incompact and slippery sandy soil, is the basic composition of sandy vegetation in grassland region. One of the biggest marks of sandy habitat is the frangibility and fluidity of matrix. Typical psammophytes possess developed root system and the ability of asexual reproduction, and have special adaptations to sand cover, sandstorm and sand flow. *Psammochloa villosa*, *Hedysarum fruticosum*, *H.mongolicum*, *Oxytropis psammocharis*, *Artemisia halodendron*, *A. wudanica*, *A.ordosica* and *A.sphaerocephala* are typical herbages and subshrub psammophytes. *Salix flavida* and *Caragana korshinskii* are the representatives of shrub psammophytes.

The above-mentioned ecological types are marked according to the correlation between plants and single ecological factors, known as single factor of ecological species group. But in the growth and development process, the plants are affected by combined effects of a great many ecological factors, presenting comprehensive and diversified adaptive functions. Plants growing in the same communityhave similar adjustment mechanism and behaviors to ecological environment. Ombrophylilous plants, which exist depending on precipitation, are able to adapt to the "common habitat" with the characteristics of good drainage, loam, medium fertility, medium pH, full solar radiation and caloric balance. For example, the dominant species such as *Stipa baicalensis*, *Filifolium sibiricum* and many forbs of *Leguminosea* and *Compositae* have adaptive convergence to meadow habitat, and *Stipa grandis*, *Stipa klemenzii*, *Artemisia frigida*, *Artemisia xerophytica krasch* and *Ajania pallasiana* etc. have similar adaptive characteristics to arid and semi-arid typical steppe and desert habitats.

Dynamics of Grassland Vegetation

Dynamic change Dynamic change usually refers to two aspects—seasonal changes and interannual changes, it is also the dynamic changes of grassland community in space-time community structure and functions.

Seasonal changes Impacted by temperate continent climate, the seasonal changes of grassland vegetation are not only reflected in "seasonal aspect", but also in seasonal difference of vegetation productivity. Due to the climate difference in different areas of temperate grassland, the seasonal changes of grassland vegetation are different everywhere. Forage grasses growing on Inner Mongolian Plateauof mid-temperate regionbegin to bud in late April, enter the growing period between June and August, and then enter into dormancy in winter. The growth period is about 150 days. Forges growing in Loess Plateau, which located in the warm-temperate region, begins to bud at the end of March or at the beginning of April, enters the best growth period between May and September, and the above-ground part becomes withered and yellow in late October. The growth period lasts 210 days.

Corresponding with the rainfall rhythm, the grassland community, which exists in the east of temperate grassland in China, only has one growth peak; but the grassland community, which distributed in the west of temperate grassland in Mediterranean climate region, has two growth peaks.

The locating observational results show that the above-ground productivity in temperate grassland community is the highest in autumn and the lowest in spring. The imbalance of seasonal productivity impacts the stable development of grassland animal husbandry.

Interannual changes In temperate grassland, as atmospheric precipitation changes greatly in different years, the productivity of grassland vegetation also changes. Observation results show that interannual productivity change of meadow steppe is smaller, about 30%; typical steppe is about 50%, and desert steppe is the highest, up to 60%~70%. According to the correlation analysis between productivity and rainfall of Hulunbuir typical steppe community, in drought years, production of 1t dry matter needs to consume 2,000t water. For example, production 100kg/mu hay needs about 300mm atmospheric rainfall. On the contrary, if the rainfall is abundant, the water consumption decreases greatly, and production of 1t dry matter only consumes 675~700t water. For example, production of 100kg/mu hay only needs 120mm rainfall. The phenomenon that the herbage production fluctuates with the rainfall is universal. The monitoring results from 1979 to 1987, which came from Inner Mongolia grassland ecosystem orientation study stage of Chinese Academy of Sciences (CAS), show that annual rainfall capacity not only restricts the above-ground biomass of *Stipa grandis* and *Aneurolepidium chinense*, but also affects the accumulation of under-ground biomass.

Thus it can be seen that the vegetation productivity of temperate grassland is mainly restricted by atmospheric rainfall. Even for the same plant community, the overground and underground biomass in drought year is only half of the biomass in rainy year. The interannual difference of productivity is different in different grassland formation. Different synusia and population have different response to climate change in the same grassland community. In a year of moderate rainfall, perennial bunch grass, xeric dwarf subshrub and deep-rooted shrub grow at a normal rate, but the growth of mesophytes and annual and biannual plants is restricted to a certain degree. But during years of abundant rainfall, the latter plants flourish, and the productivity increases greatly. During the drought year, mesophyte and xerophytes of shallow root system grow poorly, and only the deep-rooted shrubs can grow normally.

Monitoring data show that the structural characteristics (including specific composition and the relative ratio of quantitative characteristics) of grassland community, which fixed in the long-term natural selection, is relatively steady in the interannual change. The characteristics which are related to plant vegetative growth and reproduction, such as growth rate, biomass and phenophase, are decided by hydrothermal condition in that very year. The interannual change of grassland community is also affected by the climate in previous year. This effect will definitely affect the renovation and reproduction of the next year.

The succession of grassland vegetation and the transformation of ecological function The seasonal change and interannual fluctuation of grassland vegetation have some connection with the succession of grassland community. It is a long and complex process.

In nature, the succession of grassland vegetation is reflected in two directions—positive and retrogressive. Community structure and function are different along with different succession direction. Positive succession begins from the gathering of pioneer plants, through opening half-close stage-close stage, achieving the transformation from unistratal community to multiple community. The life form translates from annual and biannual herbage to perennial herbage. As time goes on, the diversity of species gradually increases through the ecological selection and habitat selection, and trends to be stable. The stratified structures of over-ground community and underground community become more complicated, achieving the coordination between the plants and the environments. Water and nutrient resources in soil are fully utilized, and change the environment at the same time, consequently, the extremity environment becomes more moderate, arid and wet environments gradually transit to secondary environment, barren soil becomes fertile, flowing sand becomesfirm, and saline-alkali soil develops to desalination. The ecological functions of grassland vegetation, including resistance to wind erosion, water erosion, decreasing the erosion of the slope, water conservation andsand fixation, become more perfect. The grassland ecosystem achieves to relative stable stage, which is known as the climax stage or mature stage of plant community.

Various kinds of *stipa* steppe are the symbol of developing to climax stage.The grassland communities formed by other populations are metastable in different stages of succession in a sense, and the structure and functions of such types of grassland communities are not perfect, or in the degraded stage of the *Stipa* influenced by grazing, mowing, fire and other external causes. Retrogressive succession makes the original mature grassland community disintegrate gradually, original ecological functionshad lost,and grassland ecosystem was destroyed finally, leading to the decreasing diversity of species and the sharp reduction of productivity. This status not only restricts the development of stockbreeding, but also leads to serious desertification.

Main Formations of Grassland Vegetation

Because of the caloric difference caused by the latitude and elevation, the grassland in China can be divided into three types—including moderate-temperate grassland, warm temperate grassland and alpine steppe. Under the influence of southeast monsoon and Mongolian high pressure, the precipitation in China's grasslands decrease by degrees, but the dryness increases from southeast to northwest (Table 3-1).

Meadow Steppe (Photographed by Xu Zhu)

Xilin Gol Typical Steppe (Photographed by Xu Zhu)

Sunit Desert Steppe Stretch to the Horizon (Photographed by Yong Shipeng)

Chapter 3 Grassland Vegetation 45

Warm-Temperate Grassland on Loess Plateau (Photographed by Wang Shunli)

Dry Steppe (Photographed by Yong Shipeng)

Alpine Steppe (Photographed by Xu Zhu)

Table 3-1 Main grassland types in China

Grassland ecological type / Formation group and formation	Moderate-temperate grassland			Warm-temperate grassland			Alpine steppe		
	Meadow steppe	Typical steppe	Desert steppe	Meadow steppe	Typical steppe	Desert steppe	Meadow steppe	Typical steppe	Desert steppe
I. Formation *Stipa* spp.									
1. Formation *Stipa baicalensis*	+								
2. Formation *Stipa grandis*		+							
3. Formation *Stipa krylovii*		+						+	
4. Formation *Stipa kirghisorum*	+	+							
5. Formation *Stipa capillata*	+	+	+						
6. Formation *Stipa bungeana*					+				
7. Formation *Stipa gobica*			+						
8. Formation *Stipa klemenzii*		+	+						
9. Formation *Stipa orientalis*			+						
10. formation *Stipa caucasica*			+						
11. Formation *Stipa stapfii*			+						
12. Formation *Stipa breviflora*			+			+			
13. formation *Stipa glareosa*			+						
14. Formation *Stipa purpurea*								+	+
15. Formation *Stipa subsessiliflora var. basiplumosa*								+	
16. Formation *Stipa subsessiliflora*								+	
II. Formation *Festuca* spp.									
1. formation *Festuca ovina*	+								
2. formation *Festuca sulcata*	+	+	+						
3. formation *Festuca kryloviana*							+		
4. formation *Festuca pseudovina*							+		
5. formation *Festuca olgae*	+						+		
III. Formation *Caespitose*									
1. Formation *Cleistogenes squarrosa*		+							
2. Formation *Cleistogenes songorica*			+						
3. Formation *Cleistogenes mucronata*					+				
4. Formation *Agropyron cristatum*		+	+						
5. formation *Aristida triseta*								+	
IV. Formation *Bothriochloa ischaemum*									
1. Formation *Bothriochloa ischaemum*					+				
V. Formation rhizome grass and rhizome Carex									
1. Formation *Aneurolepidium chinense*	+	+							
2. Formation *Orinus* spp.							+	+	
3. Formation *Carex moorcroftii*									+
VI. Formation forb									
1. Formation *Filifolium sibiricum*	+								
VII. Formation *Allium*									
1. Formation *Allium polyrrhizum*			+						
VIII. Formation *Thymus* spp.									
1. Formation *Thymus mongolicus*					+				
IX. Formation *Artemisia*									
1. Formation *Artemisia frigida*	+	+	+						
2. Formation *Artemisia gmelinii*		+			+				
3. Formation *Artemisia giraldii*					+				

Notes: This table integrates from "Vegetation of China", "Vegetation of Inner Mongolia", "Vegetation of Tibet" and "Vegetation of Xinjiang", which is not very perfect, and for your reference only.

Formation *Stipa* spp.

In China's grassland region, Formation *Stipa* spp., Formation *Festuca spp.*, Formation *Leucopoa,* Formation *Cleistogcnes squarrosa*, Formation *Agropyron cristatum*, Formation *Bothriochloa ischaemum* and Formation *A. adscensionis*, etc. form an integrated tussock-grass steppe.

Formation *Stipa Spp.* distributes in the mountainous region of semihumid, semi-arid and arid regions in the Northern Hemisphere, *Stipa* grassland community with different constructive species is an important reference index for dividing grassland regions.

There are about 300 species of *Stipa* plants around the globle, 27 species in China, of which 16 species are constructive species, composing formation *Stipa*, and the main representative formations are as follows:

1. Formation *Stipa baicalensis*

Stipa baicalensis is a kind of perennial closely-cluster grass, it is a special formation in the east

Stipa baicalensis Steppe (Photographed by Yong Shipeng)

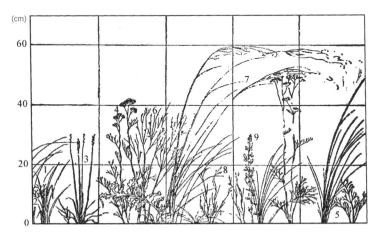

Vertical Projection of *Stipa baicalensis* (quoted from Vegetation of China)

1. *Anemarrhena asphodeloides*   2. *Artemisia desertorum*   3. *Koeleria cristata*   4. *Filifolium sibiricum*   5. *Rhaponticum uniflorum*   6. *Astragalus tenuis*
7. *Stipa baicalensis*   8. *Cleistogenes squarrosa*   9. *Lespedeza hedysaroides*

of Asia, and a representative of meadow steppe. The distribution center of formation *Stipa baicalensis* lies in the north and east of Mongolian Plateau, extends northward to Russian Yakutsk, enters to Northeast Plain passing through Greater Khingan eastward, and reaches to Changdu area of Tibet to the west.

Stipa grandis Steppe (Photographed by Xu Zhu)

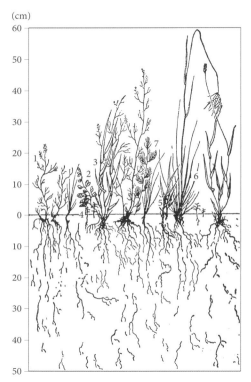

Vertical Structure Chart of *Stipa grandis*+*Aneuvolepidium chinense* (quoted from Grassland of China)

1. *Artemisia commutata*   2. *Medicago ruthenicus*   3. *Agropyron cristatum*   4. *Chenopodium aristatum*   5. *Aneurolepidium chinense*   6. *Stipa grandis*   7. *Salsola collina*

Vertical Projection of *Stipa grandis* + *Cleistogenes squarrosa* Community (quoted from Vegetation of Inner Mongolia in 1985)

1. *Stipa grandis*   2. *Cleistogenes squarrosa*   3. *Serratula centauroides*   4. *Glycyrrhiza uralensis*   5. *Bupleurum chinense*   6. *Artemisia frigida*   7. *Veronica incana*   8. *Orostachys fimbriatus*

Formation *Stipa baicalensis* is the meadow steppe with the most abundant species composition, the most complicated synusia structure and the most colorful seasonal aspect.

Formation *Stipa baicalensis*, which develops on

2. Formation *Stipa grandis*

Formation *Stipa grandis*, one of the representative formations of typical steppe, connecting with formation *Stipa baicalensis*, formation *Filifolium sibiricum* and formation *Aneurolepidium chinense*, is a specific bunch grass steppe in central Asia. Formation *Stipa grandis* never enters into desert steppe westward.

Stipa grandis is the tallest dense-cluster grass in *Stipa* plants in China, the mean height of leaf can reach 40~50cm, and fertile tillers can reach 80~100cm. The seasonal aspect of formation *Stipa grandis* is not so colorful as formation *Stipa baicalensis*, but simple and unadorned.

The floristic of formation *Stipa grandis* is less than formation *Stipa baicalensis*. It is regional, and its origin has some connections with Eurasian grassland and North American grassland.

Formation *Stipa grandis* is the symbol of typical steppe landscape, and is distributed in high plain of eastern Xilin Gol.

3. Formation *Stipa krylovii*

Formation *Stipa krylovii* and formation *Stipa grandis* are the representative formations of typical steppe. In typical steppe region, both of them are distributed crossly. Formation *Stipa grandis* has been gradually substituted by formation *Stipa krylovii* nearing the side of desert steppe, since the laterhas higher drought tolerance than the former. The species composition of formation *Stipa krylovii* is simple, reflecting the characteristics of typical steppe.

chernozem, is fit for growing spring *Triticeae* crop, and is the reclamation object for constructing crop and forage bases.

Stipa krylovii+Artemisia frigida Dry Steppe (Photographed by Yong Shipeng)

Formation *Stipa krylovii* is the most fundamental pasture in the north of China, and over-grazing and reclamation may easily induce land desertification.

4. Formation *Stipa capillata*

Formation *Stipa capillata* is the most extensively distributedin Eurasian grassland. It starts from European grassland, gets across Ural Mountain, extends to west Siberia and Kazakstan, reaches to Altay of Xinjiang, north slope of Tianshan Mountain and west mountainous region of Junggar, and enters to Mongolian Plateau int he east, where it is substituted by formation *Stipa krylovii* and formation *Stipa grandis*. Formation *Stipa capillata* grows in different grassland region and forms different ecological variation types.

5. Formation *Stipa bungeana*

Stipa bungeana, also known as *Stipa bungeana,* is a kind of warm-like bunch grass in China. Formation *Stipa bungeana was* once widely distributedin Loess Plateau in the midstream of the Yellow River, but now, a large area of *Stipa bungeana* grassland disappeared due to thousands years of agricultural reclamation, and it is difficult to find closely grouped *Stipa bungeana* grassland. Formation *Stipa bungeana* is the main body of warm-temperate grassland, its distribution centers are in Loess Plateau of Shanxi, Shaanxi, east of Gansu and south of Ningxia.

The species composition of formation *Stipa bungeana* is simple. The reproductive habit of *Stipa bungeana* includes seed propagation and bulblet propagation. After autumn rains, many bulblets will be generated at the basal of leaf sheath, separate from the parent body, drop into the soil, and then form new seedlings. Because the reclamation of *Stipa bungeana* is too long, resuming the aspect of primary community is impossible.

6. Formation *Stipa* breviflora

Formation *Stipa breviflora*is the representative formation of warm temperate desert steppe in China. It is distributed in the transition location of typical and desert steppes, the Loess Plateau of the midstream of the Yellow River, belonging to agro-pastoral transitional zone.

The specific composition of formation *Stipa breviflora* is simple.

Now, large-area *Stipa breviflova* grasslands have been reclaimed and dry farming has been developed agriculture, but the yield is very poor and instable, the phenomenon of desertification is very outstanding.

7. Formation *Stipa glareosa*

Formation *Stipa glareosa* is an important desert steppe formation in the middle of Asia (including Qinghai-Tibet Plateau grassland region), the eastern and northern boundaries of distribution area are roughly consistent with formation *Stipa gobica*, but the western and northern boundaries go beyond the distribution area of *Stipa gobica*. In China, formation *Stipa glareosa* is distributed in the northwest of Inner Mongolia, but the area is very small. *Stipa glareosa* has very strong adaptability to arid climate in desert steppe.

It is an important desert steppe type in the northern mountainous region of Xinjiang and Ali area of western Qinghai-Tibet Plateau.

8. Formation *Stipa purpurea*

Formation *Stipa purpurea* is the most representative formation in alpinesteppe. It occupies the center of Qinghai-Tibet Plateau—Qiangtang High Plateau, lakes and basins in southern Tibet, upper middle vales and alps in Brahmaputra, western Qinghai Plateau and high mountains of central Asia (Qilian Mountain, Kunlun Mountain, Tianshan Mountain and Pamirs Plateau).*Stipa purpurea* is the proper bunch grass in Qinghai-Tibet Plateau, Pamirs Plateau and central Asia alps, and has strong cold-resistant and drought tolerance characteristics.

Formation *Stipa purpurea* is a good type of pasture with high quality.

9. Formation *Stipa gobica*

Formation *Stipa gobica* is the most basic small-sized bunch grass steppe in medium moderate temperate desert steppe of central Asia. Formation *Stipa gobica is* distributed in Wulanchabu Plateau and midwest areas of Erdos in China. It is one of the most drought-tolerant formations. *Stipa gobica* is a con-

Formation *Stipa gobica* (Photographed by Yong Shipeng)

Landscape of Formation *Stipa gobica* (Photographed by Yong Shipeng)

Stipa klemenzii (Photographed by Yong Shipeng)

structive species with strong drought resistance, and it is not only the constructive species of desert steppe, but also the main constructive species of zonal desert steppe vegetation.

In addition, the original source of the *Stipa gobica* is also permeated by eastern Mongolian grassland composition and xingan-Mongolian components (such as *Caragana microphylla*, *Astragalus galactites*, *Stipa krylovii Roshev* etc.), Western Asia and Kazakhstan, Mongolia components (such as *Caragana pygmaea*, *Convolvulus ammannii*, etc.) and the Alxa Desert components (such as red sand), and contains the floristic elements in northern temperate zone (such as *Eragrostis minor*, *carex duriuscula*), old World temperate (such as *Ptilotricum canescens*, *Thymus quinquecostatus* and *Artemisia frigida*.) and ancient Mediterranean (such as *Kochiaprostrata*) and other flora. This zone is characterized by the results of plant combinations due to inland arid climate changes in the long historical process.

Formation *Stipa gobica* is the most widely distributed short-grass pasture in desert steppe. The quality of *Stipa gobica* is very good, fitting for ingesting by sheep, goats, horses and camels.

Formation *Festuca* spp.

Formation *Festuca ovina* is distributed in temperate grassland region and tropical mountainous regions, it is the second largest grassland resource on land, only inferior to formation *Stipa*. There are about 100 species of *Festuca* worldwide, and 23 species have been found in China, extensively distributed in Northeast of China, Inner Mongolia, Xinjiang, Tibet, Qinghai, Gansu, Sichuan and Yunnan, etc., but its community ecological range is relatively narrow, with no large-sized continuous distribution.

Formation *Festuca ovina* often appears in the top of low mountains and hills and the top of windward slope. The formation *Festuca ovina* distributed in low mountains and hills in western Manzhouli of western Hulunbuir grassland is the only one which distributes concentratedly. Formation *Festuca ovina* only appears in the southeast of Qiangtang High Plateau in Qinghai-Tibet Plateau, where the elevation is 4,900~5,100m.

Besides, formation *Festuca ovina* also appears in mountainous regions in desert and sub-alpine in subtropical forest regions.

Formation *Bothriochloa ischaemum*

Formation *Bothriochloa ischaemum* is a representative type of warm-temperate forest steppe, distributed in low mountains and hills in the west and north of north China and the middle and south of Loess Plateau.

Bothriochloa ischaemum is a perennial bunch grass, originating from tropic, processing short rhizome and strong asexual propagation, wear tolerance and resisting erosion. *Bothriochloa ischaemum* is not only a type of forage, but also a soil-conserving plant.

Formation *Cleistogcnes squarrosa*

Formation *Cleistogcnes squarrosa* belongs to shortgrass steppe. There are 15 species around the globe, and 11 species in China. *Cleistogcnes squarrosa* is a xerophic bunch grass, distributing in Eurasian grassland region.

In China, formation *Cleistogcnes squarrosa is* mainly distributed in typical steppe region in Inner Mongolian Plateau and sporadically distributes in Erdos High Plateau and the west of Songliao Plain. Formation *Cleistogcnes squarrosa* grassland can be divided into two types, one is formed by over-grazing, and the other is formed in coated sand covering region.

Formation *Agropyron cristatum*

Agropyron cristatum is a xerophic bunch grass, widely distributed in Eurasian grassland region. In China, formation *Agropyron cristatum* is often found in sandy land of grassland region. The area of formation *Agropyron cristatum* is very small and often unsteady. Therefore, the grazing pressure must be controlled in order to avoid desertification of grassland.

Agropyron cristatum grows mixing with *Koeleria cristata*, *Poa sp.* and *Cleistogcnes squarrosa*, forming short-grass steppe, which is a transition type in the

Cleistogenes squarrosa Steppe Photographed by Yong Shipeng

stage of succession.

Formation *Aristida triseta*

Formation *Aristida triseta* is distributed in tropic arid region in the world. There are 150 species all over the world, and 8 species in China, of which, *Aristida triseta* is a known community constructive species, constituting Formation *Aristida triseta*.

Formation *Aristida triseta* distributes in the north of Tibetan Himalayas, the south of Nyainqentanglha Mountain and the middle reaches of Brahmaputra, which is often found in the hillsides and the foot of mountains with an elevation of 3,700~4,000m.

A. adscensionis is a common plant inmoderate temperate and warm temperate grassland region, considered as the symbol of grassland degradation.

Formation *Aneurolepidium chinense*

Formation *Aneurolepidium chinense* is the proper rhizome grass steppe in the Central Asia sub-region of Eurasian grassland region. It extensively distributes in the east of Inner Mongolian Plateau and Songnen Plain. *Aneurolepidium chinense* has wide ecological range, belongs to mesophytes or wide-xeri plants, with the resistant capacity of salt tolerance and waterlogging tolerance. Formation *Aneurolepidium chinense* is a type of grass steppe with the highest plant communities and the most complicated species composition in China's grassland region. It is the priceless natural fortune in China's grassland resources, processing the highest yield of forage in temperate grassland.

Formation *Aneuvolepidium chinense* (Photographed by Xu Zhu)

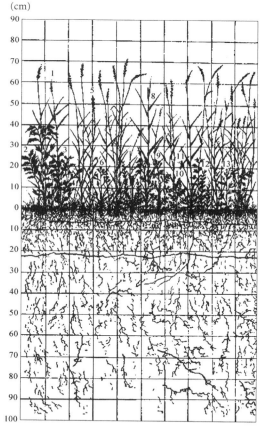

Vertical Projection of *Aneurolepidium chinense* + formation forb community (Songnen Plain) (quoted from Vegetation of China)

1. *Aneurolepidium chinense* 2. *Serratula centauroides* 3. *Filifolium sibiricum* 4. *crested hairgrass* 5. *Calamagrostis epigejos* 6. *Lxeridium gracile* 7. *Gueldenstaedtia verna* 8. *Phragmites australis* 9. *Saposhnikovia divaricata* 10. *Glycyrrhiza uralensis* 11. *Scorzonera austriaca* 12. *Heteropappus altaicus* 13. *Polygonum divaricatum*

Formation *Orinus kokonorica* and Form. *O. thoroldii*

Formation *Orinus kokonorica* and Form. *O. thoroldii* are the grassland vegetation communities mainly composed by *Orinus kokonorica* and *Orinus thoroldii*. It is the proper formation in Qinghai-Tibet Plateau, often distributes below the elevation of 4,600m. Formation *Orinus kokonorica* and Form. *O. thoroldii* possess good functions of soil and water conservation. There are many withered stem leaves existing in this type of grassland, it is yellowish-green in summer. The quality of the grassland community is inferior, the palatability is poor and the productivity is low.

Formation *Carex moorcroftii*

Carex moorcroftii is the proper floristic element in Qinghai-Tibet Plateau. Formation *Carex moorcroftii*, mainly composed by *Carex moorcroftii*, distributes in the north of Qiangtang High Plateau over a large area, also appears in the west of Ali and the source of Brahmaputra, where the elevation is 4,900~5,400m. Formation *Carex moorcroftii* is the grassland resources distributed in the highest location with regional significance, laying a foundation for the development of alpine steppe ecosystem. Formation *Carex moorcroftii* is the main pasture in northern Tibet Plateau, providing the basic forage sources for Tibetan-livestock and local wild animal, although its stem is hard, palatability and production is poor.

Formation *Filifolium sibiricum*

Formation *Filifolium sibiricum* is a typical representative of forbs steppe in China's grassland. The grassland community, composed by *Filifolium sibiricum* (as the constructive species), is the proper formation in mountainous and hilly areas of east Eurasian grasslandregion. The distribution range is approximately between 100°~132° E and 37°~54° N. In general, the distribution range of formation *Filifolium sibiricum* is roughly consistent with the natural boundary of forest steppe. Professionals usually take the appearance of formation *Filifolium sibiricum* as the symbol of entering into meadow steppe.

Filifolium sibiricum is the "god of longevity" in herbage plants. According to the scientific research,

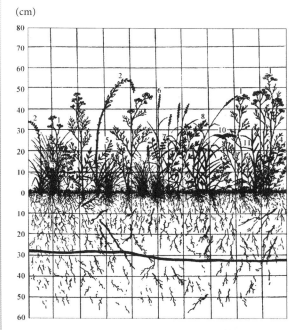

Vertical Projection of *Filifolium sibiricum* Steppe Community (Songnen Plain) (quoted from Vegetation of China)

1. *Filifolium sibiricum* 2. *Rhizoma anemarrhenae* 3. *Stipa baicalensis* 4. *Saposhnikovia divaricata* 5. *Scutellaria baicalensis* 6. *Aneurolepidium chinense* 7. *polygala tenuifolia* 8. *Lespedeza hedysaroides* 9. *Potentilla flagellaris* 10. *Spodiopogon sibiricus* 11. *Cleistogenes squarrosa*

the most longevous *Filifolium sibiricum* can live more than 130 years, this phenomenon is rare in herbage plants. Besides, *Filifolium sibiricum* is a kind of "late childbirth" plant, which grows very slow, and blossoms and yields fruit for the first time at the age of 15~20. The species composition of formation *Filifolium sibiricum* is similar with that of formation *Stipa baicalensis*, which reflects their consistency in origin. Formation *Filifolium sibiricum* is the most beautiful and changeful grassland in China's grasslands.

Formation *Artemisia frigida*

Formation *Artemisia frigida* is the grassland community which takes *Artemisia* (xerophic subshrub plant) as the main body with a certain amount of other grasses. Many *Artemisia* plants are distributed in Chinese grassland, in which *Artemisia frigida* occupies the largest area.

Formation *Artemisia frigida* is widely distributed in Mongolian Plateau and Ordos Plateau, existing as the constant companion of formation *Stipa* and formation *Aneurolepidium chinense*. The large area

Artemisia frigida (Photographed by Yong Shipeng)

of formation *Artemisia frigida*, which is found in the grassland, is not the original community, but is the result of degradation caused by intensive grazing and strong erosion on formation *Stipa* and formation *Aneurolepidium Chinense*. It is a succession forma of grass steppe, the floristic composition is relatively miscellaneous, representing the trend of drought in the degraded grassland.

Artemisia frigida is favored by sheep and other animals, no matter fresh or withered, this is an obvious difference from another odorous *Artemisia* plants.

Artemisia frigida has the characteristics of turning green in early spring and withering in late winter, so *Artemisia frigida* can solve the problems of feeding animals under the condition of extreme scarcity of forage grass and feed.

Formation *Thymus* spp.

Formation *Thymus spp.* is a peculiar grassland type composed by many

Artemisia frigida Degraded Grassland (Photographed by Yong Shipeng)

Thymus plants and steppe grasses. The *Thymus spp.* is native to temperate regions in Europe, North Africa and Asia. In China, it is mainly distributed in western Liaohe River Basin and in loess hilly region of the middle Yellow River, and becomes a representative type with landscape meaning in this region. Besides, formation *Thymus spp.* occupies a certain area on the fixed sandy land (such as Maowusu sandy land, Kubuqi desert and Xiaotengeli sandy land, etc.) in the grassland region. Formation *Thymus spp.* mainly distributes in typical steppezone, and also extends to meadow steppe, but generally does not enter into desert steppe zone. From originating sources, formation *Thymus spp.* can be divided into two types, one is original, often grows in extreme habitat, such as skeletonsoil and cretaceous outcrop, the other is the retrogressive succession stage of original grass steppe due to extensive grazing and erosion. In nature, the later phenomenon is more universal.

*Thymus spp.*is a dwarf subshrub with volatile bioactive compounds and smells sweet. It is not only a kind of high-quality, but also the raw material for producing essence and condiment, and is called as "ground Chinese prickly ash" by herdsmen.

Formation *Allium polyrrhizum*

Allium grassland refers to a type of herbage plant community mainly composed by *Allium*. There are 450 species of *Allium* around the world, and 120 species in China. Most species are mesophyte vegetation, appearing as auxiliary species in the community. The grassland which takes *Allium* as the constructive species is known as *Allium polgrrhizum* desert steppe.

Formation *Allium polgrrhizum* extensively distributes in the Mongolian Plateau, appears from Hulunbuir Plateau in the east and enters desert steppe to Junggar Basin, its distribution center is in desert steppe region. Formation *Allium polgrrhizum* occupies large area of salinized brown calcic soil, and becomes the concentrated distribution area of *Allium* desert steppe.

Allium polgrrhizum is a kind of typical super-xeric bunch bulbaceous plant, having strong grazing tolerance ability. The seasonal aspect of formation *Allium polgrrhizum* often fluctuates with the inter-annual precipitation and the coming time of rain season. Commonly, *Allium polgrrhizum* germinates in early spring when it snows more in winter or has some rainfall in spring. Generally, itturns green in April, blossoms in

Formation *Thymus spp.* during Full-blossom Period (Photographed by Yong Shipeng)

Formation *Allium polyrrhizum* (Photographed by Xu Zhu)

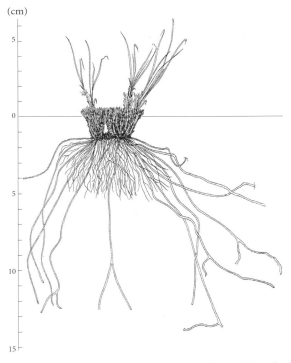

Tussock and Root System of *Allium polyrhizum* (quoted from On the Biological Characteristics of *Allium polyrhizum* in 1977)

July, and ripes in August. But if the weather is severely drought in spring, *Allium polgrrhizum* stays in "die-away" state for a long time, and only grows a few short and tender leaves. Naturally, the productivity of formation *Allium polgrrhizum* is very low, often less than 50g/m^2 of fresh grass, subordinate to low yield grassland.

Allium polgrrhizum is a valuable forage resource in desert steppe, it has high nutritional value. When grazing sheep and camels in autumn in *Allium polgrrhizum* grassland, the quality of meat and milk will be improved, the oestrus of livestock can be promoted, and the reproduction rate of animal will be improved. Besides, formation *Allium polgrrhizum* is the best location to collect *Nostoc Flagelliform*, an edible blue-green algae of state-protected plant.

There are 176 grassland formations in China, including 14 meadow steppe formations, 24 typical steppe formations, 32 desert steppe formations and 14 alpine steppe formations. The 13 formations introduced here are just part of the representative formations, from which we can see the complexity and diversity of Chinese grassland vegetation formations.

Chapter 4
Grassland Animal

The Sun rises on the Prairie
Written and composed by Meili Qige

In the blue sky float the white clouds,
Under which the horses gallop.
Crack my whip, sending the sound to everywhere,
And flocks of birds begin to fly up.
If somebody comes to ask me,
What place is this?
I'll tell him with pride,
This is my homeland.
The people here love the peace,
And love their homeland.
We sing about our new life,
And an ode to the Communist Party.
Oh, Chairman Mao and the Party,
You have nurtured our growth.
The everlasting sun is rising in the Prairie.

The ecological position of the grassland animals

In solar system we call the earth "biological planet". The surface layer automatic control system of the earth is composed of biosphere, atmosphere, hydrosphere and lithosphere. The network composed of complicated food chain plays an important role in the system.

Animal is the accumulator and transformer of energy in the surface layerautomatic control system of the earth. The large animals on grassland such as *Cervidae, Bovidae and Equidae* consume a lot of plants everyday. Rodents are small in size but they have high propagate ability and a large number of populations. They play an important role in substance and energy transformation. *Insecta* is the successful resident on grassland. They are primary consumer and secondary producer and main food resources of many birds and beasts. Carnivorous animals such as *Felidae, Canidae, Mustelidae* and *Strigiformes,* and birds such as *owls and Falconiformes* are top consumers, some kinds of them such as *Hyaena, Aepyginae* search cadavers with sharp sense of smell, which accelerates the circulation of materials.

With the existence and sacrifice of animals, grassland ecosystem can keep homeostasis between energy grade and the number of animal by the rule of life. For example, the decline in number of wolf can result in the overflow of rodents and grassland degradation; the degraded grassland results in the decline of reproduction or increased death rate of *Rodentia,* which keep the rational level of population quantity. For instance, the arctic lemming can lay 12 young per litter, but if the number is too much, the color of their hair will become reddish orange, sending a warning to lower reproduction rate; if the number increases, some individuals may even jump into the sea to suicide to prevent the species from extinction by sacrificing themselves.

Each animal occupies a position in the nature by following the nature according to genetic code with morphologic, physiologic andecologic characteristics as "calling card", and plays an important role in nature.

Coevolution of grassland animals and grassland environment

In this topic we try to use fossil to explain the evolution of grassland and grassland animals. It is very difficult, but the development of biology is a teaching book of modern biology with dynamic characteristics, making sense for explaining the coevolution of the grassland animal and grassland environment.

The Mesozoic phytophagous animals and predators

In the early Mesozoic, about two hundred million years ago, continents on the earth linked together and known as "Pangaea". There were many fern, gymnosperm and herbage at that time, the terrain was basin, and the weather was warm and humid. Except *Arthropoda*, *Vertebrate* and ancient reptiles developed quickly and gave birth to many branches which can adapt to any environment. One of them is original mammal, which is called *Multituberculata*, *Prototheria* mammal such as *Ornithorhynchus anatinus* dwelling in Australia may be related to *Multituberculata*. *Multituberculata* is the only type of mammal feeding on grass in earlier period. Although they are original and followed the pattern of oviparous, they grow furs and breast, which feed their offsprings by milk. *Metatheria* animal is marsupial which may be born in America in Mesozoic. With the bloom of angiosperm, they left for Europe, Asia, Africa, and dwelled in Australia in Cenozoic tertiary metaphase and got great development since then. This kind of animal is viviparous but the placenta is very original. Their cubs were given birth during the embryonic period. Even after tens of millions of years of evolution, the modern *Macropus giganteus* with a length of 2 m have their young with body length of only 2.5 cm, just like the size of red dates. With the developed forefoot, the young climbs to brood pouch of its mother to eat latex. When they hold nipple in the mouth, the nipple bulks up and muscle contract and then the milk flows into the young's mouth. After 7~8 month growth in the brood pouch, the young can live independently. Another branch of the early Mesozoic ancient reptile turned into dinosaurs which grow quickly in lake basin condition with many plants. They overpower original mammals and have many branches which adapt to various living environment, such as huge animals eating plants and other feral animals eating meat. They died out dramatically in the late cretaceous period after over 1.6 hundred million years' reign on earth.

During the Pangaea period, the animal composition around the world was similar and had not too much differentiation. Until the Eocene epoch of Neozoic the connected continents were flooded by seawater, and the Pangaea divided into current continents, which gradually formed the animal composition of different continents.

Herbivores and carnivores in Neozoic Vegetarian

65 million years ago, the earth history entered into Cenozoic era including Tertiary Period and Quaternary Period. The Tertiary Period is subdivided into Paleocene, Eocene, Oligocene, Miocene and Pliocene epochs. The Quaternary Period is subdivided into Pleistocene and Holocene. The angiosperm got a bloom period in Cenozoic. The *Insecta* and placenta mammals which have perfect modes of reproduction have made an unprecedented development and tachytely.

1. Herbivores in the Paleogene era

Paleocene, Eocene and Oligocene, which were 65~23 million years ago, are called early Paleogene. At that time, Paleocene and Eocene have stable geographical conditions; Eurasia was not connected with Africa and India peninsula, and Tethys was between them. There were abundant rainfall and rivers all over the world and climate was warm and suitable. Many arbors occupied the main position, the land was covered with forest, evergreen broad-leaf forests grewin plain, and Coniferous forest spread all over the mountains. The fossil data showed that it was mainly forest landscape in North America 40 million years ago. The forest have gestated ancestor of grassland animal. The huge hoofed animal is the leader of grassland animal and their ancestor was called *Condylarthra* which appeared in the forest of early Paleocene. *Tetralaenodon* may be included the ancestor of *Perissodactyla* such as horse and rhinoceros. They are similar to *Hyracotherium*. *Hyracotherium* is only as large as a fox which lived in North American forest and eats leaf. *Mesohippus* appeared in Oligocene and was as big as a sheep and their live environment changed to grassland. At the same time, other large-size *Perissodactyla* such as *Brontotheriidae* developed well and lived in wetland. 13 generic animals such as *Embolotherium grangeri*, *Proeitanotherium mongoliens* lived in Inner Mongolia. The ancestor of *Artiodactyla* such as cow, sheep, antelope, pig, elephant and camel appeared later than *Perissodactyla*. In early Eocene, the original animal was *Diacodexis Gazin* which was very small and stayed in forest. The ancestor of deer, *Archaeomeryx*, which has no horn and is only as big as a rabbit, lived in Asia. The ancestor of elephants evolved from African forest and the relative ancestor of modern elephant is *Moeritherium*, which has no long nose and is as large as a pig.

The grassland carnivorous vertebrates include various kinds of *Felidae*, *Canidae*, *Mustelidae* and raptor which evolved nearly at the same time with herbivores. The earliest carnivorous mammal which was called *Miacidae* appeared in late Paleocene, and entered into prosperous period in late Eocene and early Oligocene. *Metailurus* is the ancestor of *Canidae* and *Felidae*.

Because of plate movement, the Himalayan orogeny started in the early Eocene epoch. At that time, Qinghai-Tibet Plateau, the Himalayas, Alps in Europe, Rocky Mountains in North America, and Andes in South America ascended continuously. The windward side of each mountain range formed the terrain rainfall along with the rising of humid airflow caming from ocean, but the leeside of mountain range is very dry and temperature is higher with the decrease of height. The precipitation decreased because vapor can not be coagulated so that a large area of dry region appeared in the lee-side of mountain range and near areas. In remote areas the precipitation and temperature increased because of the change of general atmospheric circulation. At the same time, the orogenic movement made Tethys separate from Tibet and only residual part remained.

The ocean current changed and prevented the change of energy which affected climate.

In the early Oligocene, the global climate began to become cold and the continental climate appeared in subtropical zone and variable zone. Many factors have contributed to the appearance of patchy meadow steppe and grassland in the middle of North America, the middle of Eurasia, south of South America, south of Africa, Sahara Desert, highland and plain of the western China. The ancestors of grassland animals walked out of forests or moved in ground. The expanded grassland, large lake basin and marsh provided platforms for grassland animal to develop. Some kinds changed their feeding habits and added more grass in food. The lagging development of carnivorous mammals let the herbivores live longer on grassland and their size is bigger and bigger, such as the *Megatherium americanum* which is 1t weight, and the *Dinornis* is 3 m in height lived in South America. The Perissodactyla, which evolved firstly, became a common species on grassland and their size changes from small to "huge", e. g. 13 kinds of *Brontothere* lived in Inner Mongolia. In addition, *Chalicotheidae, Tapiroidae, Amynodontidae, Indricotheriidae* have also appeared, and *Artiodactyla* such as *Gobihyus* appearedalso. The offspring of *Archaeomeryx* changed their digestive system in grassland environment.Their esophagus became big and formed stomachus compositeswhich can ruminate. They developed quickly spreading from Asia to North America in late Oligocene because of lacking of enemy. Except for mammal, the *Eogrudae* and *Struthio camelus* fit for open ground began to flourish.

The development of herbivores offered a variety of foods for predators. Carnivores began to develop in late Eocene and early Oligocene. Some species hunted insect in early time and then changed to hunt mammal. Take the evolution of dog as an example, from *Cynodictis* in late Eocene to *Pseudocynodictis* in Oligocene and then to *Cynodesmus* in Miocene Epoch. The old civet cat which is the ancestor of *Felidae*, civet cat, *Hyaena* appeared in late Eocene. The *Felidae* became a single species which is soft but powerful predator.

2. The synergetic development of grassland environment and grassland animal in Neocene period

In the process of coevolution of grassland environment and grassland animal, the Neogene period is an important transition period because the geology, climate, vegetation and animal evolved during that time.

The movement of Himalayas made the height of Himalayas, Alps, and coastal mountain chains in North and South Americas ascended sharply, and Qinghai-Tibet Plateau rose to 2,200~3,000m above sea level. The High mountains and plateaus have changed atmospheric circulation, making the inside of the continent drier than before. The continental drift changed the geographic latitude. According to calculation of geomagnetism, Eurasia and Africa drifted northward and China drifted northward about 1,100~1,400 m from Eocene to now and the temperature decreased along with the increased latitude. At the same time, from Miocene Epoch, the weather became colder starting from North Pole and South Pole. The icecap of South Pole formed in Miocene Epoch and the icecap of North Pole formed in Pliocene. The northern boundary of sub-tropics moved to 35 degrees north latitude and formed the dry and cold climate. Forest landscape has been replaced by grassland,and the "king of grassland"-*Stipa* appeared at the same time.

Upland meadow descended to plain and formed vast grassland. Grassland became the main type of terrestrial ecosystem so that the grassland animal developed well and had a good community succession. In the Eogene, some specific large herbivores declined due to climate change, *Brontotheriidae*, *Amynodontidae* and *Artiodactyla* such as ancient pigs have extinguished in succession. In the Neogene period,a cousin of *Amynodontidae* developed into many kinds of *Rhinocerotidae* such as *Chilotherium hoberei and Chilotherium sp*, etc.

The *Moeritherium* which has long nose and fore-tooth can adapt to forest and grassland environment. They can eat food with fore-tooth and long nose, and they eat tall grasses also, so that they developed very quickly and expanded to all continents except Australia. There have been 400 kinds of *Moeritherium* successively. With the changing of food and gradual increase of medium-sized elephants, they have changed the environment and played an important role in the transformation from forest to retama bushland.

With the expanding of grassland, many artiodactylas in the Miocene Epoch such as cow, sheep, Gnu, yak and ariel have appeared one after another on grassland. The deer and camel have further developed and many of them reduced the chance of being attached by enemies through rumination and fought against predators with their horns. But the exception is that the giraffes have flocky horns and never change. They eat leaves and then evolved to special body figure that have long neck and high height. The giraffes use big hoof to battle with other animals and have a share on grassland. *Palaeotragus tungurensis* which came from Inner Mongolia is a member of it.

The evolvement of horse is a good evidence for the coevolution of grassland environment and animals. Horses originate from North America. The hyracothere in early time looked like as big as a fox and lived in forest but big as a sheep in Oligocene. They have three toes in front and hind leg sand their middle toes became longer. They lived in the transitional zone from forest to grassland. In the Miocene Epoch, the forest in temperate zone was gradually replaced by grassland. There are two important branches of *Miohippus*. One is *Mesohippus* with bigger body size and degraded toes and another is *Merychippus* which only eat green grasses.The *Merychippus* has heightened corona and their surface enamel got into deep layer of dentine, forming a mazy structure with dentine, which increased the abrasion resistance and grinding ability of teeth. The ancient horses expanded into many varieties such as *Pliohippus*, *Hipparion* and *Equus* in modern era with the development of grassland. They expanded to Eurasia and South America, and the *Hipparion* died out until the late Pleistocene. Another branch of *Miohippus* likes to eat green leaf and expanded to Eurasia, such as *Anchitherium gobiensis*, *Sinohippus zitteli*, *Sinohippus sp*, etc., but all of them died out until the late Neogene period.

At the same time, the carnivores have developed with the evolution of herbivores. The grassland is not good for herbivores to hide but provides opportunities for hunting by carnivores. The *Machairodus* which have big clumsy body and moves slowly lived in Eurasia and America, some kinds of which have about 20 cm sharp teeth. The *Hyaena* of *Canidae* uses powerful neck, sharp teeth and sensitive nose to battle with huge *Felidae* for food.

3. The Pleistocene in the Quaternary Period is the fast succession period of grassland animals

The Quaternary Period includes Pleistocene and Recent Epoch, is the last century of Cenozoic. The Pleistocene lasted from about 2 million to 10 thousand years ago, and the Holocene lasted from ten thousand years to now. Important events happened in Pleistocene including the earth's continental displacement. The mid-latitude regions of Asia, North America and South America were located at the westerlies. The Himalayan movement was intensified, ice age and interglacial stage appeared alternately, and mammal evolved quickly and humankind entered the historical arena.

The Himalayan movement in Quaternary Period made the mountains of Europe, Asia, and America ascended rapidly. The Himalayas has ascended over 3,000 m in Pleistocene. The Alps in Europe, Mountains of Scandinavian Peninsula, Rocky Mountains in North America, Andes in South America and Himalayas stopped the wet air mass of ocean which came from Atlantic Ocean, Pacific Ocean and Indian Ocean. The rain shadow resion of leeside formed rain shadow desert, and adjacent inland formed the climate suitable for grassland, and the temperate zone in South America and North America formed a large area of temperate grassland. The continental drift affected ocean current which resulted in the transition from thedry season to rainy season. The woodland steppe appeared in Africa, South America and Australia due to water shortage. The airtemperature decreased in ice age and the climate was dry. The original forest region under dry and cold condition was replaced by grassland and developed quickly in Africa, Eurasia, Central Asia, Med, and south central of North America. A research confirmed that the desert in the northwest of India and Sahara were distributed with dotted grassland in the ice age. The high temperature and drought during the interglacial period made the range of grassland worldwide further enlarged.

In a word, the earth movement during the Pleistocene epoch, the alternation of ice age and interglacial stage, the succession of tundra, forest, grassland and desert made animals experienced nature's power of selection. Therefore, the Pleistocene is the great development period for animals' extinction, migration and rapid development. The evolvement of elephants was a good example of this. *Trilophodon* was the ancestor in the Pliocene epoch, *Mammuthus* and mastodon appeared in Pleistocene. When ice age was coming, the elephants were forced to hold back in the south and evolved to *Elephas maximus* and *Loxodonta Africana*.

Though *Mammuthus* resists cold with long furs and eats grass with curly fore-teeth, they disappeared finally in late Pleistocene or early Recent Epoch about 8,000 years ago. *Stegomastodon*, *Mastodon americanus* appeared in the end of Pliocene but disappeared in early Pleistocene. The extinction of large-scale herbivores made some huge carnivores came to an end of theor evolution. For example, the *Machairodus* which was more than 300kg disappeared because of the extinction of big herbivores, furthermore, they had no ability to battle with fast-running preys and *Felidae*. *Acinonyx jubatus*, hyena, rhinoceros andgiraffe which were distributed in China in the Neogene period also migrated to the south to South Asia and even Africa.

The great changes of ecological environment in Pleistocene, the accumulation of genetic material of grassland animals as well as the complication of gene pool also provided opportunities for the birth and development of new species. Most important genera and species of modern mammals appeared at that time, such as artiodactyla including *Bovidae, Antilocapridae, Ciraffidae, Cervida,*Perissodactyla including *Equidae, Tapiridae, Rhinocerotidae* and other main species including *Felidae, Canidae, Mustelidae, Hyaenidae* and *Viverridae*, etc.

4. The Recent Epoch is golden ages of modern grassland animals

When the Recent Epoch entered into the late ice age, from 10 thousand years ago to today, no obvious changes have taken place about the distribution of sea and land over the planet's surface. Though climate fluctuated in a short period, the vegetation has no big changes in large areas. The grassland area in the world was 24,000,000km^2 which accounted for 1/6 of total land area. Grassland animals are similar to those of Pleistocene, such as *Equus hemionus*, *Equus. Przewalskii*, *Camelus knoblochi*, *Gazella subguutturosa*, *Ovis ammon* and *Canis lupus*and *Panthera spp*, which have been on earth from Pleistocene to this day. But in the history of evolution of the Recent Epoch, the factors that affect animal life have impacted on human activity. The human ancestors were not powerful and they ate animal flesh raw and drank its blood in the early time, have the same position with animals. With the development of human beings and increasing of intelligence and wisdom, they have gradually become the master of other cerateures' life, and grassland animal community was not solely controlled by nature.

With the increasing pressure on the use of natural resources and ecological environment, the global desertification is accelerated, and the area of desert and sand is extending. We humans have been using technology to intervene with natural evolution and to squeeze the basic living space and food resources of animals to the greatest extent, and many animals were even driven out. For example, 3,000 years ago, the Elephas maximus lived in Shanxi province, China, but now they were forced to stay in a corner of Yunnan Province. Some animals had a more miserable life and died out finally under double pressure of nature and human, for example, woolly rhinoceros and mammuthus distributed in Asia thoroughly stepped down from the stage of history.

The *Rodentia* can contend with human and reproduce quickly and have many kinds of life styles, thus, they adapt to many conditions. The *Muridae* developed in early Cenozoic but now it becomes the vermin which is very difficult to deal with. The grassland degradation is harmful for huge grassland animal, but the rats can see their enemies and run away easier than before because of the short vegetation. Currently, there are more than 1,700 types of rats recorded all over the world.

As a result of human's interference, the bloom and extermination of animals were not only controlled by the law of nature. The order of grassland ecosystem has been broken by illogical use of human beings. If we ask for everything without control and take nature lightly, the self-regulation system on earth surface will accommodate by itself if the ecosystem has no order, at that time, maybe a small pathogenic microorganism and other nature disasters will punish human rigorously because of their activities of violating the scientific laws.

Distribution of Grassland Animals in the World

Grassland animal communities have their general and specific characteristics all over the world. There are some historical reasons for this distribution of them, above all the difference of modern ecological environment.

The continental drift is the reason for separating and combining of each continent that affected the distribution and diffusivity of animals. Today, the grassland animal distribution has direct connection with plate movement. As previously mentioned, about two hundred million years ago in late Paleozoic and early Mesozoic, continents on the earth linked together and were known as "Pangaea", which had nearly the same animal community. Dinosaur was the leader of animals and inferior mammals and aves appeared at that time, such as *Ornithorhynchidae* and *Tachyglossidae*. From late Mesozoic, the Pangaea began to split, animals with similar gene began to evolve in different environment and finally formed different eco-geography communities. According to the research on land fauna for over a hundred years, we divide fauna into 5 geographic areas as follows:

Marsupial mammals of Australia

In the late Mesozoic, when Australia and Sorth America separated from the Pangaea, they became an absolute island continent. The marsupial in the Cenozoic era of passed South America and flowed to Australia, and then Australia separated itself from South America and never connected with other continents, therefore, the placental *Eutheria* risen and developed from North America and Eurasia did not enter Australia because of the separation of ocean. The marsupial has few placental enemies and Australia became their refuge, they stayed and developed there. Today, there are 16 families, 254 species of marsupial all over the world, among which, 14 families are distributed in Australia, New Zealand and nearby islands, and half of which are grassland animals. *Macropodidae* is typical steppe animal which is only 0.5kg in weight and has different size and shape, but *Macropus qiganteus* which is 100kg in weight and 2 m in height, is a typical steppe animal on open grassland.

The marsupial entered Australia with ancestor's gene. With the same ancestor's gene and similar environment, the marsupial could have parallel evolution with placental mammals, and give birth to many animals which have similar shapes and behaviors as placenta such as *Dasyuridae, Thylacinidae, Notoryctidae* and *Macropodidae*, etc.

Kangaroo

Particular grassland animal community of South America

After separating from Pangaea in late Mesozoic, Marsupialia such as *Nootoungulata, Thoatherium, Toxodon* and another species developed in South America. At the end of Tertiary, the South America continent connected with North America and landbridge appeared between them. Many kinds of placental *Eutheria* entered South America but the hoofed animals declined quickly, and completely died out with the invasion of many huge animals like *Machairodus* until Pleistocene. The *Didelphidae* is the most famous *Machairodus* survived in South America. The female *Didelphidae* pouch has 19 nipples and has strong reproductive ability. When their children get out of pouches then lived on mothers' backs with the tail enlaced, and this is the reason for their name. Some specialized species of *Machairodus* such as *Caenolestidae* have survived today. Some animals came from North America, such as *Lama guanicoe* and *Vicugna vicugna* are grassland runners and live together, share the same gene with camel. The *Cervidae* of hoofed animal included *Blastocoros, Pudu* and *Odocioleus bezoarticus*, etc. *Myrmecophaga tridactyla* and *Dasypodidae* belong to *Edentata*, which are original placental mammals distributed only in Central America and South America and only eat insects because of their vestigial teeth. *Myrmecophaga tridactyla* lived closed to the land. They are 2.4 m in length and have long and thin head and nose. Their vermiform long tongues can flex skillfully, and they also have big and drooping tails, powerful forelimbs and powerful claws. The young aborigines here take the challenge with *Myrmecophaga tridactyla* as a sign of a real man. *Dasypodidae* has about 20 species with armor on body and lives in grassland, forest and other environment. The research approved that the *Dasypodidae* is highly resistant to ieprosy bacillus, which provides good clue and material for preventing and controlling lepriasis, thus the *Dasypodidae* is protected by people. There are 8 kinds of *Dusicyon* in carnivores and *Puma concolor*, *Panthera onca* which entered America in Pleistocene were on the top of nutrition grade. Aves have their own characteristics in South America. It is well known that *Trochilidae* is the smallest bird in the world and the smallest one is only 0.001kg in weight. In fact, the family of *Trochilidae* is very big and the number comes to 329 in total which distributed in America. The South America is the place where there are 254 kinds of *Trochilidae*, accounting for 78% of all. *Trochilidae* is the only bird that can fly backwards with beautiful flower feather and metal luster. They can get orientation exactly and suck nectar by their proboscis. *Rheiformes* including *Rhea Americana* and *Rhea pennata* are special species of South Africa.

Grassland animals of Africa

Africa with a land area of 30,000,000 km^2 includes the largest area of tropical savanna in the world. The southern Africa has temperate grasslands. It is well known that Africa is the kingdom of grassland animals. Because Africa was separated from Eurasia and connected with land bridges many times, animal communicated with each other frequently and the common grassland animal appeared, including *Perissodactyla* such as *Equidae*, *Rhinocerotidae*, and *Artiodactyla* such as *Asartiodactyla Tragelaphinae*, *Bovinae*, *Hippotraginae*, *Antilopinae*, which were the common subfamily of Asia, Eurasia, North America and Africa.

However, due to earth rotation, Coriolis Effect has been formed, and the northern Africa and the Arabian Peninsula located in subtropical high belt with high air density and dry air. The desert was formed with dry and cold weather in ice age, and with the obstruction of Sahara and Arabian deserts, the grassland in central and southern Africa grassland became an isolated island. After long-time evolution, the grassland animal community in Africa has great speciality.

Gerenuk

Connochaetes (Photographed by J.M. Suttie)

Giraffa camelopardalis (Photographed by J.M. Suttie)

Hippopotamus amphibius (Photographed by J.M. Sutt)

anthera leo and bubalus carabanesis (Photographed by J.M. Suttie)

Dicerorhinus (Photographed by J.M. Suttie)

Equus burchellii (Photographed by Xu Zhu)

Loxodonta (Photographed by J.M. Suttie)

Bovidae is a big type of huge herbivores which has 9 subfamilies and 4 of them are unique species of Africa, such as *Cephalophinae, Neotraginae, Alcelaphinae* and *Reduncinae*, and more than 50 kinds of antelopes including 7 species of *Grazella, Connochaetes gnou, C. Taurinus* and *Tragelaphus oryx,* weighing up to 600kg weight, *T. strepsiceros* with 168 cm length of big horns, *Cephalophinae* which is only 85 cm in height, and *Bubalus bubalus.*

Hippopotamus amphibius is an amphibious animal. With its oil substance from sebaceous gland, they can prowl or swim in water area to avoid enemy and torridity. At night, they often look for green grass and fruit in sparse steppe far away from water area. Other endemic species including *Giraffa camelopardalis, Ceratothcrium simum, Diceros bicornis, Equus burchrlli, E. grevyi* and *E. zebra, Loxodnta, aardwolf, Canis mesomelas, Lycaon pictus, Panthera leo* and *Phacochoerus aethiopicus,* etc. In a short distance, the speed of *Acinonyx jubatus* when hunting is 110 km/h, but if a lot of heat energy in the body can not be released as soon as possible, their nervous system will be hurt by the increasing temperature in the body, so the time of preying on must be in one minute, otherwise they have to give it up and the success rate is only 50%. Only the claw of *Acinonyx jubatus* of *Felidae* can not flex, so they can keep balance with the attrition of claw. The *Mandrillus sphinx* which has red nose and blue face, the *Papio cynocephalus* and *Erythrocebus patas* of *Cercopithecidae* also live in African grassland. *Struthio camelus* is also an endemic species of African grassland.

Grassland Animals of Eurasia and North America

The reasons for the similar characteristics of grassland animal composition in Eurasia and North America are the consolidations and divisions of lands in history as well as the similar characteristics of climate and geographic position. During the early Neogene period, the North Atlantic Ocean was not connected with the Arctic Ocean, North America was connected with Europe, many placental mammals including the ancestors of grassland hoofed animals and carnivorous fossil were discovered in Europe and America and there were many common families and genera in animals between them. In the Eocene epoch in mid-tertiary, Europe was separated from North America, but animal communities also have the same gene at that time. Meanwhile, due to the dry up of the Ob River, Eurasia connected with each other and then turned into the biggest plate tectonics. The ice age of the Quaternary Period made sea level declined, land bridges appeared many times between Eurasia and North America which provided a passageway for animals to communicate. Its only 13,000~14,000 years after the separation of Eurasia and North America, thus it has a long communication history of animals and there were various common species such as *Cervus elaphus, Capra ibex,* wolf and *Vulpes,* etc. Each animal evolved from them can mix together, for example, horses and camels evolved from North America extended to Eurasia and *Felidae* evolved from Eurasia extended to North America, therefore, when dividing geographic zoning of animals, Eurasia and North America were incorporated into Holarctic realm. Of course, the expanding of animals was obstructed by climate, for example, *Felidae* entered into North America until Pleistocene and *Giraffa camelopardalis* have never set foot on North America because of the cold weather.

Though Eurasia and North America connected with each other many times, the difference of land form, physiognomy and climate resulted in the difference of the eco-environment. Animals lived in the isolated special conditions evolved into many species such as *Antilocapra americana*, *Ovibos moschatus*, *Bos bison*, *Canis latrans* and 13 kinds of *Mephitis mephitis* which are endemic species of North America. *Saiga tatarica*, *Procapra gutturose*, *Grazella subgutturosa*, *Equus hemionus*, *Equus ferus*, *Camelus bactrianus*, *Pantholops hodgsoni* and *Bos mutus*, which are endemic species of Tibet plateau. *Felis bieti* and *Felis manul* are only distributed in Asia. *Rupicapra rupicapra*, *Bison bonasus*, and *Mustela putorius* are endemic species of Europe. According to the historical reasons for animals, we take comprehensive analysis on the similarities and differences of modern animal community, the holarctic realm is classified into the palearctic realm of Eurasia and Nearctic realm of North America.

Relative to the kingdom of grassland animal in Africa, Eurasia and North America do not have so many species of animals as Africa because of high latitude and weather conditions. Especially, although the species and number of ariel are limited, the ariel has many rare species, which are worthy of investigation.

Grassland Animals of Oriental Realm

Oriental realm includes the India Peninsula to the south of the Himalayas, Malaya Peninsula, Qinling of China and south of Huaihe River. Most of the original vegetation were forest, only the northwest and south central of India Peninsula have savanna, and some azonal grassland distributed in other areas. Until late Mesozoic, the India Peninsula connected with Africa and entirely united with Asia in the Eocene epoch of early Cenozoic about 45 million years ago through continental drift. Therefore, the fauna of Oriental realm has the same species as Africa, for example, the *Elephas maximus* are distributed in India but their relative species is African elephant. There are 5 families of *Rhinocerotidae* in the world, *Rhinoceros unicornis*, *R.sondaicus* and *Dicerorhinus sumatrensis* were distributed in Oriental realm, while *Ceratotherium simum* and *Diceros bicornis* were distributed in Africa. *Panthera leo* were distributed in Africa, Oriental realm and northern India. In the process of evolution,

Antilocapra americana (Photographed by J.M. Suttie)

Bos gaurus (Photographed by J.M. Suttie)

the Oriental realm has many kinds of species such as *Cervus porcinus, C. unicolor, C. eldi, Boselaphus tragocamelus,* and *Tetracerus guadricotnis,* etc.

Grassland Animals of China

The grassland area of China is about 2,000,000 km^2, accounting for 1/5 of total land area. There are various types of grassland, including moderate-temperate grassland, warm-temperate grassland, alpine steppe and upland meadow. With the different history and environment, each grassland has different grassland animal communities and the animals can adapt to various environment with their special abilities and skills.

The Mammal

There are 430 kinds of mammals distributed in China. At least half of mammals are typical steppe animals or animals whose living environment is related to grassland, among which, there are herbivores, carnivores and other omnivorous mammals.

1. Herbivores

We take *Equidae, Bovidae, Ceridae, Camelidae* and *Rodentia* as the typical species of herbivorous mammals in China. *Equidae* belongs to *Perissodactyla*. China has 3 kinds of them including *Equus ferus, Equus. hemionus* and *E. kiomg,* which have the common characteristics of the family— extended limbs, only the third toe and the horniness of hoof which can grow and wearproof.

The spring ligament which is composed of elasticin is between phalanx and metacarpal bone of the *Equidae*. When they lift their feet, the ligament will shrink and hoofs rebound automatically, when they put their feet down the ligament will tense, so that the friction and the forward push will be increased at that time, but the flex of hoofs does not need the shrink of muscle and energy, thus they can run a long way on grassland. The equids have no horns as weapons of defence, but they can use their powerful hind limbs when they face enemies. They are vigilant animals whose neck looks like a thick and strong spring, while eating, they will raise their heads frequently to see surroundings. According to the working principle of spring, the equids can save as much energy as possible, it is recorded that one horse may raise head 1,000 times per day.

Bovidae, Ceridae and *Camelidae* are artiodactyla. There are 15 kinds of bovine whose environment

Kiang (Photographed by J. Marc Foggin)

Naemorhedus goral (Photographed by Yang Guisheng)   *Pseudois nayaur* (Photographed by Xing Lianlian)   *Cervus elaphus* (Photographed by Xing Lianlian)

Camels on Desert Grassland (Photographed by Xu Zhu)

is connected with grassland in China, including *Bos gaurus*, *B. mutus*, *Grazella subgutturosa*, *Procapra gutturosa*, *P. picticaudata*, *P. przewalskii*, *Saiga tatarica*, *Pantholops hodgsoni*, *Naemorhaedus goral*, *N. cranbrooki*, *Capricornis sumatraensis*, *Budorcas taxicolor*, *Capra ibex Pseudois nayaur* and *Ovis ammon*. The Ceridae include *Cervus albrostris*, *C.elaphus*, *C. unicolor* and *Capreolus capreolus*, etc. *Camelus bactrianus* is the only kind of *Camelidae* in China. The artiodactylas mentioned above are ruminant, their stomaches are composed of chambers forming stomachus composites, the front 2~3 chambers can reserve, intenerate and ferment food, and they can store up food in stomach in a short time and then ruminate the food to their mouth in a safe place, then eat it slowly, digest and absorb it absolutely. Rumination can decrease the chance of eating by other predators.

Many artiodactylas have long limbs and their metacarpal bone develops into the "cannon bone" which connected with talus and formed the flexible arthrosis, therefore, the sheep, deers and oryx have fast-running and jumping abilities. Their long neck can move flexibly and extend the range of collecting information of surroundings. They are observant and alert and extremely vigilant. Once they find a sigh of disturbance or trouble, they can show a clean pair of heels. These kinds of animals have powerful arms—the horn, which is used to battle for mating right and resist the enemy. The horn of chiru like the edge of knife can cut through a wolf's abdomen. Cattles, sheep and oryx have bone horn which wears a horniness sheath and never changes and branches off. The horn of deer changes once a year and increases one furcation each time until the furcation reaches a certain amount. For example, a red deer has another name -"eight—horn deer" for eight furcations. *Capreolus capreolus* has three furcations at best. When a deer changes horn and has a new horn, the new horn is called "cartialgenous" because it has ossified shaft with multiple foramen and blood vessel and very soft velours skin. After aging, the outer skin falls off and becomes the bone horn.

Camelus bactrianus lives in desert steppes and desert areas. They have very good sense of smell which is good for searching water source;the stomach of *Camelus bactrianus* has three chambers, the first and second chambers have many water cells which are constituted by inwall ruga. Their red cells can expand to absorb water so that camel is able to endure drought summer and increase body temperature by 1~2℃, moreover, camels have strong ability to endure dehydration and they can also live under the condition with 27% dehydration. Camels are called "ships of desert" and are the main means of conveyance in harsh desert climates. They have soft and flat feet that are covered by 1cm thick protective pads, which form the bottom of their feet with 2~4cm thick elastic tissue. This configuration formed the big and soft feet of camels, which help them work freely in the desert. Camels' feet have lots of melanin, which can prevent the heat energy of desert from being transferred to camels. When the surface temperature reaches at 70℃, camels can maintain their nomal physiological function in a long time just with a little salt to keep the supply of inorganic salt ion in their body.

Rodentia is a major class group of grassland animals in herbivores. It has a big family and the variety and number are superior to others in mammal. There are 200 kinds of *Rodentias* in China, and about 100 kinds of which are grassland muroid, which plays an

Grass stored by Ochotona alpina (Photographed by Xing Lianlian)

important role in grassland eco-system. The murine is not only the primary consumer but also the secondary producer. *Mustelidae*, *Felidae*, *Canidae* and serpentes take them as primary food sources, so *Rodentia* is an important part of energy flow and circulation of materials in earth's biosphere. Except for the *Marmota* of *Rodentia* with bodylength of 60 cm on grassland, most of other animals are small. They can hibernate, store food, live in caves to avoid harsh climate, they take many methods such as living together, using the sensitive organ and giving an alarm to hide and seek with their enemies. Because of their higher innate capacity for increase, they can keep the advantage of community quantity. Many rats on grassland live in caves, using shady and wet condition to decrease water dissipation of skins and lungs. The rats have different caves because of different species and soil texture. *Citellus dauricus* like hard soil grassland and their caves have few holes even only one hole, but the underground construction is complicated which includes soft couch grass bedroom and toilet. It is comfortable and safe. *Petauristidae* often live in sand and move at night. They have piercing eyes which can receive much more light and fetch food at night. They dig holes at dawn and disperse wet sand like a fan which leaves no trace at sunrise. There is a branch near the path of main cave and is called clearstory which is less than 10 cm away from the ground; if the cave was discovered by enemies, they can run away through this clearstory. *Gerbils* live in soft soil and their caves have many holes including temporary holes and living holes.

Temporary holes are very simple to prevent sunstroke and high from enemies, but living hole is complicated. This kind of rats does no hibernate and use food stores to pass the winter. Their caves include bedroom, toilet and some warehouses. Different winter foods are stored in different storages and the process is very clear such as bean storage, grain storage and so on. If they have no time to produce in crop place, they will bring some stover to cave and produce them carefully and then ship the stover out from caves. *Meriones unguiculatus* can store food as many as 25kg. During the famine when farmers have not enough food or meet disasters, they will dig rat caves to rob food. *Ochotona daurica* has another name ——*Pallas*, they often store green grass in autumn and pile up grass such as *Artemisia* near the entrance to cave. They will unfold grass and dry it in the sun to prevent going moldy. *Dipodidae* and *Citellus dauricus* hibernate in holes in late autumn and do not need to eat, store food and drink. Their body temperature will drop to 0.7~5 ℃ in hibernation. They breathe a few times and the oxygen consumption accounts for 1/40 of nomal value. *Marmotas* breathe 2~3 times/min in hibernation and they can live until next spring with a small amount of energy consumption.

Allactaga sibirica

Meiiones unguiculataus (Photographed by Yang Guisheng)

Humankind take their own benefits as the standard to weigh the pros and cons of *Muridae*, but we do not know that these secondary producers sacrifice themselves to feed many tropical level animals such as fox, wolf, raptor and snake to maintain the order of grassland ecosystem. The overrun of rats is the result of competition between humans and animals. The overuse of noxious pesticide and indiscriminately arrest and hunt of raptors and mammals decrease the number of rodents, which has lower increasing speed than rats. This gives chance for rats to increase the quantity of community without the control of enemies. The grassland degradation caused by overgrazing, the decreasing height of plants and reduced cover degree of grassland provide open vision for rats to evade oncoming danger and enter into holes. We human beings consider rats as dreadful monster, but we should think deeply about the phylosophy of balance between humans and animals. We should think over our activities with scientific attitude because any live beings have a reason for existence.

2. The Carnivores

Of China's grassland carnivores, *Felidae*, *Canidae* and *Mustelidae* have the largest number which can hunt herbivores by their perfect physical structure and super hunting skills. With the awl-shaped canine, the carnivorous mammals can cut in neck or other critical parts of enemies and kill them quickly without wasting body energy in battle. Their carnassial can tear food like forfex, with which they can prey on food easily on grassland.

Felidae includes some important species such as *Felis silvestris*, *F. bangalensis*, *F. bieti*, *F. manul* and *F. lynx*, etc. Their body is soft and they can increase their steps as their vertebral column works like the spring. The speed of *Felidae* is amazing but their endurance is low, so that they are secret killers of nature and often ambuscade other preys. In order to avoid being found out by preys, many kinds of *Felidae* have special claws which can flex. When they go forward, the muscles in the back of toe flex and draw the claw. They touch down to the ground voicelessly and catch the quarry with the two-side muscle of toe shrinking at one time and extending claws as soon as possible.

Canidaes are fierce killers of grassland with not so many speciessuch as *Canis lupus*, *V. corsac*, *V. ferrilata*, and *Cuon alpinus,* etc. Their number is large and they move astutely. It is often said that foxes appear and disappear in quick succession, which shows their characteristics of alert and quick. Wolves are good at hunting and also work collectively. The egregious endurance and speed are the condition for their success. Mongolia horse is the Perissodactyla which has the best endurance. This is the result of being chased by wolves for a long time. This is the rule of living in animal world.

Mustelidae is common species of grassland animals, including *Mustela sibirica*, *M. altaica*, *M. eversmanni*, *M. erminea*, *M. nivalis*, *Martes foina*, *Meles meles* and *Arctonyx collaris,* etc. Many of them are long and thin but ferocious and have strong viability. When they meet enemies, they may hurt themselves in order to run away. The author once caught a fitchew with a mouse trap; it tried to run away by biting the nipped legs. There is a pair of scent gland beside the anus of *Mustelidae* and the secretion of it can be used to sign their field and attract opposite sex, but at the crises of hunting by enemies, the scent gland will spurt some very old mash which are unbearable by any hunters, and when enemy is frightened, they can run away as soon as possible. The skull of *Mustelidae* is very weighty and shoulders can shrink to core axis. Their body can pass though caves if their head can enter into caves, and they can break into caves to kill rats.

The activity time of animals is logical in order to reduce competition in the evolvement process. Some animals work at night and others work in daytime and their activity rhythm is different. Many kinds of *Felidae* and *Canidae* and their quarry such as deer, oryx are nocturnal animals. In order to adapt to the weak light condition, the rod cell in retina increases twice as much the diurnal animals. In the back of retina, there is a thin and smooth crystal theca named "shine theca". When the light shoots into the retina, the shine theca reflexes the light and stimulate the retina again which strengthenes the light. In the moonlight, the blue and green light in wolf's eyes looks like the moving stars which can shake in preys' shoes.

Grassland Birds

Of 1,250 kinds of birds in China, there are about 400 kinds of summer migratory birds, passing birds and winter migratory birds dwelling in grassland, among which, 200 kinds of birds reproduce in grassland and the green grasslands have fostered many kinds of birds which adapt to the grassland environment. The herbage is constructive species of grasslands, and many of them use their developed fibre to sink rainwater and snow like sponge. Though the grassland is dry, the grass seed also can have a good harvest and offer birds enough food. Dry and soft earth's surface is good for hatching of grasshopper. *Coccinella septempunctata Linnaeu*, *Tettigoniidae*, grasshopper, ant, locust, ants and butterfly can be seen everywhere on grassland. Birds often take insects as good meals. Some birds which have nests in ground are the main part of grassland, such as *Alaudidae*,

Mustela eversmanii (Photographed by Li Xiaohui)

Pteroclididae, Phasianiae, Charadriidae and *Turdidae,* etc. Large birds such as *Gruidae, Anthropoides virgo, Grus vipio, Otididae, Otis tarda, Chlamydotis undulate* and *Tetrax tetrax,* etc. *Accipitridae* is in the top trophic level of grassland, including *Buteo hemilasius, B. buteo, Aguila rapax, A. chrysaetos, Circus cyaneus, C. macrourus, C. spilonotus,* and *Falconidae* such as *Falco tinnunculus, F. vespertinus, Strigidae* such as *Bthene noctura, Athene noctua* and *Asio otus*,etc.

There are 13 kinds of *Alaudidae* in grassland of China, among them, *Melanocorypha mongolica, M. maxima, Calandrella cheleensis* and *Galerida cristata* are the resident birds which never migrate over long distances and stay in their homeplace all the year around. *Calandrella cinerea, C. acutirostris, Mirafra javanica, Alauda arvensis* and *A. gulgula* are the summer migratory birds which propagate in China in spring but migrate to another place in late autumn. *Eremophila alpestris* is summer migratory bird in temperate grassland but a kind of resident birds in Qinghai-Tibet Plateau and Xinjiang. *Melanocorypha yelteniensis* is a straggler appeared in Xinjiang by chance. *M. bimaculata* and *M. leucoptera* are winter migratory birds which pass the winter in Xinjiang grassland. The *Alaudidae* is a singer of grassland and when they want to attract opposite sex, they will sing a happy song. *Alaudidaes* build their nests close to ground and build delivery rooms in brushwood or a small hole which leave behind trample of cattle, and take dung and grass as signs.

Emberizidae is an important member of grassland birds. There are 29 kinds of bunting in China and many of them are relative to grassland environment. The familiar species include *Emberiza cioides, E. cia, E. leucocephala, E. pusilla, E. aureola, E. pallasi* and *E. koslowi,* etc. They nest bowl-like nests in grass or shrub and disguise skillfully, they fall to the ground 100 meter far away from their nests and snake to nests, thus the enemy can not find them easily.

Falco amurensis (Photographed by Pan Yanqiu)

Pteroclididaes are grassland dwellers with medium-sized body. There are 3 kinds of *Pteroclididae* in China, including *Syrrhaptes paradoxus, S. tibetanus* and *S. orientalis.* Their fly rapidly and flexibly, which are good for searching water source and escaping enemies. *Pteroclididae* often nests on dry ground and they would absorb water with quill-coverts in belly and feed yound birds.

Saxicola spp and *Oenanthe* of *Emberizidae* often dig holes or use rat holes to give birth to child. They are well known as sharing caves with rats. *O. isabellina* will even drag the young rat out of cave and occupy it to

Grus vipio (Photographed by Xing Lianlian)   *Aquila nipalensis* (Photographed by Xing Lianlian)   *Asio otus* (Photographed by Li Xin)

propagate. *Riparia riparias* often search cliff and trench wall of grassland and use their small sharp claws to dig complicated nest. There is a thickly dotted entrance to the cave which looks like the crenels of strafe. The passage way in the cave is complicated and winding, and a corner which is 30 cm far away from the entrance is their real delivery room which is covered with soft grass. Not only a bird of prey can hardly break in this nest, but even the weasel is difficult to turn around.

Otididae and *Anthropoides virgo* build simple ground nests in Achnatherum splendens and tall grasses located at the deep of grassland. When they are in danger, they will use "ostrichism"—lower head, extend neck and hide themselves in grass. They have a good command of this trick when they are young, but with the degradation of grassland, they often fail to do that because the short grass can not hide their buttocks.

Bush-wood steppe and sparse steppe increase the stratification of landscapes and enrich the composition of grassland birds. For example, some kinds of *Laniidae* (build nest in the pile of *Nitraria tangutorum Bobr*, and with the protection of *Nitraria tangutorum Bobr*, a bird of prey cannot get close to the nest. Some big birds such as *Bthene noctura* often use bushwood to cover themselves and have a rest. Sparse steppe often appears on the transitional zone between grassland and forest, and sand. There are Hunshandake Sandy Land, Khorchin Sandy Land, Hailar Sandy Land and Mu Us Desert in Inner Mongolia. Sandy land has a good ability for water penetration and water-holding, and there is no surface flow in rainfall. Sandy land has good water conditions. The sandy sparse steppes composed by arbor, shrub and grass have different birds with different space and food resource. *Sturnidae* and *Paridae* use treeholes to propagate. *Asio otus*, *Falconidae* and *Milvus Korschun* build their nests in crotches of trees. *Phylloswcopus* and *Uragus sibiricus* use bushwood, and the *Phasianidae* w including *Phasianus colchicus*, *Perdix dauuricae* and *Coturnix coturnix* use grass to build their nests. Except for that, many incomplete rock hills, highland or bluff left in grassland due to the rise and fall of stratum in the geological period, the volcano explosion and the overflow of fulgurite. *Aguila rapax*, *Aguila clanga*, *A. chrysaetos*, *Bthene noctura*, *Athene noctua* and other small *Passeriformes* such as *Tarsiger cyanurus*, *Phoenicurus auroreus*, and *Monticola spp*, etc. build their nests

Lanius isabellinus (Photographed by Pan Yanqiu)

Otis tarda (Photographed by Li Xiaohui)

Anthropoides virgo (Photographed by Yang Guisheng)

in bluff and rock slot.

The beautiful grassland landscape includes not only boundless grasslands, but also criss-cross rivers, bright and clear lakes and green wet meadows. Qinghai-Tibet Plateau has the most marshes of China. The area of inland lake is more than 21,000km² in Tibet and there are more than 300 lakes with an area of more than 5 km². Qinghai Lake is a tectonic lake with an area of more than 4,500km² in Qinghai Province, known as "Western Sea" by ancient people. Chaerhan Salt Lake was called "lagoon" which was an ancient sea left after the apophysis of Qinghai-Tibet Plateau. Dalai Lake with an area of 2,300km² and Dalainuoer Lake with an area of 230km² are the tectonic lake and barrier lake formed with the uneven sedimentation of

Birding of Anthropoides virgo (Photographed by Yang Guisheng)

stratum; and Ulansuhai Nur with an area of 293km² is a furiotile lake formed by river migration. Another kind of lake, namely, sandy lake is peculiar to grassland. Sand lake can store much water and the fountain gather together and form sandy lake group in low-ly-

Phasianus colchicus (Photographed by Pan Yanqiu)

Bubo bubo (Photographed by Xing Lianlian)

Athene noctua (Photographed by Yang Guisheng)

ing place. There are oasises and meadows around any kind of lakes.

The river, lake and other marsh contain abundant biological resources including *phytoplankton* such as *Bacillariophyta, Cyanophyta Chorophyta,* etc., *Zooplankton* such as *Cladocera, Benthos* such as *Lymnaea, Anodonta, Odonata* and*Culicidae,* etc. The food fostered various kinds and a number of fish, *Acipenseridae, Salmonidae, Cyprinidae* and *Cobitidae* are the most common. Many kinds of insects, frogs and hoptoads are prosperous in wet meadows and the grassland marsh becomes the fairyland of water birds because of the rich food resource.

Marsh is a part of important landscape of ecosystem. Some transition appeared between marsh animal community and grassland animal community. The nest area and feeding land of marsh animals and grassland animals present the interlaced spatial pattern. There are many big and small riverways and lakes in Bayanbulak grassland of Xinjiang, a famous swan lake is located here with an area of about 1,200km^2. Qinghai Lake has a mid-lake island with an area of 0.8km^2, tens of thousands of *Anser indicus* perching there. Hulun Lake and Buir Lake are located in the beautiful Hulunbeir grassland of Inner Mongolia. *Cygnus Cygnus, Anser cygnoides, C. anser, Anas falcate* and *A. platyrhynchos* swim in these lakes. *Grus japonensis,* white-naped crane and large-scale wading birds breed offsprings in the bulrush reeds and look for food in marshes and meadows. *Otis tarda,* and *Anthropoides virgo* move in grassland far away from water and look for insects, frogs, lizards or rats. *Charadriiformes* and other small wadersoften move near waterside and eat benthonic animals with their proboscis. *Larus ridibundus, Chlidonias* and *Sterna* fly over the marsh, and when they find fish, they will dive to water and prey on it as soon as possible. *Chlidonias* can fly close to the ground and catch insects on grasslands. At the time of breeding and before migrating, many kinds of gulls look for high protein food such as insects to accumulate energy. *Aguila rapax* and *Aguila clanga* like standing at the high place and stare at rats and rabbits that play on the ground. *Circus cyaneus*and *C. spilonotus* often fly in the sky and want to catch *Panurus biarmicus* and the young birds in the bulrush.

The depth of Wuliangsu Lake in the western of Inner Mongolia is less than 1 m and has strong photosynthesis because the bottom of water can absorb sunshine. There are abundant aquatic organism, which provide rich food and good hiding conditions for birds. Hundreds of *Cygnus olor*, thousands of *Platalea leucorodia*, ten thousands of *Netta rufina* and *Aythya nyroca* propagate here. Wuliangsu Lake is the biggest breeding place of *Cygnus olor*. It is a really swan lake.

Dalai Nur is located on the beautiful Gongger Grassland in Chifeng City of Inner Mongolia. There are four rivers originated from Daxinanling Mountains and Hunshadake Sandy Land. Some sandy spring on the bank of southern lake has clear spring water and feeds into the lake. The entrance of river and the bank side of spring constitute a large area of marsh which breed 200 kinds of bird, and 100 kinds of which

Chapter 4　Grassland Animal　**83**

Bucephala clangula (Photographed by Xing Lianlian)

Wetland Flock (Photographed by Li Xiaohui)

Phoenicurus auroreus (Photographed by Xing Lianlian)

Flying Grus japonensis

Limosa limosa (Photographed by Xing Lianlian)

Rostratula benghalensis (Photographed by Yang Guisheng)

breed here, such as *Grus japonensis*, *G. vipio*, *Anthropoides virgo*, *Anser cygnoides*, *Cygnus*, *Larus argentatus* and *Ardeola bacchus*. Dalai Nur is an important stage for birds' migration. It is recorded that there are 50,000~60,000 of *Cygnus Cygnus*, 20,000~30,000 of *Anser cygnoides*, thousands of *Anthropoides virgo* and hundreds of *G. vipio* stay here in mid October.

The southeastern Ordos of the Inner Mongolia and Loess Plateau of northern Shaanxi Province are covered with Mu Us Sandland, the zonal vegetation of

Glareola maldivarum (Photographed by Pan Yanqiu)

Charadrius asiaticus (Photographed by Xing Lianlian)

Panurus biarmicus (Photographed by Pan Yanqiu)

which is typical steppe and the sand vegetation is sparse steppe. There are many deflation lakes and sand lakes in the low-lying place of grassland and about 8,000 *Larus relictus* breed in the mid-lake island of Taoli Temple, Alashan Nur. This bird population is one of the biggest populations all over the world and the number accounts for 60% of total species around the globe. Therefore, it has caused worldwide concern and was listed as international important marsh in 2002.

The Huihe River marsh which located on Hulunbuir grassland of Inner Mongolia is a truly "crane garden". Over 60 million years ago, the riverway of Huihe River was a lake, so their riverbed is flat and wide with reeds spreading all over, and the environment is suitable for *Gruidae* to reproduce. The author recorded 102 red-crowned cranes one day in August 2005, and some of them took the young to look for food. The author also found

Cygnus olor (Photographed by Xing Lianlian)

Platalea leucorodia (Photographed by Xing Lianlian)

Netta rufina (Photographed by Xing Lianlian)

Grus vipio, *Grus leucogeranus*, *Grus monacha*, and other valuable species such as *Phalaropus fulicarius* and *Podiceps grisegena*, etc.

Migratory birds account for a big part of birds in China. The birds will migrate to south before the approaching of harsh winter.

Grassland Insects

If you have a trip on grassland in summer or autumn, you will find many insects, which fly in the sky, or run on the land. There are more than 100 kinds of common grasshoppers of *Orthoptera* on China's grassland, such as *Locusta migratoria* in temperate grassland, *Gomphocerus turkestanicus* in Xinjiang, *G. tihatanus*, *Gryllotalpa* and *Acrida chinensis* on Tibetan alpine steppe. The colorful butterfly and phalaenae belong to *Lepidoptera*, which has hundreds of species. The shape of butterfly antenna looks like a wooden club and their wings stand uprightly on the back. The antenna of phalaenae looks like a comb but when they stop fly, their wings are unfolding and nutant in a roof shape. Most of them move at night. The *Coleoptera* of insects includes scarab, *Anoplophora* and *Rodolia*, among them, the *Geotrupes stercorarius* is commonly known as dung beetle. It is recorded that there are about 90 kinds of *Geotrupes stercorarius* all over the world, they have very good sense of smell which can smell dung one kilometer far away from them and gather together in a short time when the new faeces appear and then cut it to small pieces like a walnut. They stand upside down and push the dung to their caves with hind legs, they often lay eggs in dung and use it to breed offsprings. The mosquitos and flies of *Diptera* have close relationship with people. There are about 100 kinds of *Aedes*, *Anopheles*, *Culex* and *Tapanidae* in China and they are really bloodsucker, which bring troubles to people and other animals.

Most insects are the primary consumer and secondary producer of grassland ecosystem, and they are the major source of food of flogs, hoptoads, cabrites, snakes, birds, and mammals, and they translate plant foods into animal protein nutrient substance which not only breed animals but also offer the edible source of animal protein to people. However, an

Larus relictus (Photographed by Yang Guisheng)

The Eco-geographical Fauna of Grassland Animals in China

China is endowed with vast grasslands. The same grassland environment evolved many same species. Therefore, in the geographic zoning of animals around the world, the grasslands of China is within the scope of Palaearctic realm. Because of the different history, topography, climate, vegetation and modern natural environmental condition of grasslands from place to place, the grasslands of China have the special members which can adapt to different life conditions, and have formed characteristic animal community, therefore they have been divided into northeast area, Inner Mongolia–Xinjiang area, north china area and Qinghai–Tibet area. But different grassland animal community has different charactereristics.

1. Moderate Temperate Grassland Animals

The moderate temperate grassland zone covers a vast geographic area, extending from the Northeast Plain, West Liaohe River Plain, Hulunbuir Grassland to the southwest and to the northern foot of the Yinshan Mountain. The north and northeast of it is connected with outside Baikal Grassland of Mongolia and Moscow, it is an important part of Asia central grassland. The hydrothermal condition of moderate temperate grassland has changed with zones from northeast to southwest and has formed the forest steppe, typical steppe and desert steppe in proper order. The animal communities in different grasslands are not the same and divided into Songliao Plain area of northeast and eastern grassland area of Inner Mongolia–Xinji-

Ardeola bacchus (Photographed by Li Xiaohui)

unavoidable problem is that insects have brought quite a few damages to human by computing for food and infectious diseases. People take insects as pest based on self-interest, but whether the insects can cause disaster or not is decided by the number. If we keep balance of ecosystem, many insects can be controlled by their enemies at a reasonable level. Unfortunately, the improper economic activities of human disturbed the order of the nature, moreover, the decreasing number of raptors also give insects a chance to run away. Humans must follow the law of nature, preserve the ecological environment and let the law of nature limit harmful insects, Preserve them from becoming the big disaster.

Perdix dauuricae (Photographed by Pan Yanqiu)

ang region. Because Daxinganling Mountains, north Hebei mountain area and the Yinshan Mountains are located in this grassland and under the influence of mountain climate, the grassland and meadow steppe appeared, accordingly, the animals which lived in sand steppe and meadow steppe appeared in the grassland animal community.

The forest steppe is in the west foot of Daxinganling Mountains and the foot of southeast of mountain. The forest and grassland vegetation coexist. The climate belongs to moderate temperate with annual precipitation of 400mm. The vegetation grows well, the cover-condition and food condition for animals are good. The mammals such as *Capreolus capreolus*, *Nyctereutes procyonoides*, *Vulpes uulpes*, *Marmota sibirica*, *Myospalax psilurus*, *Myospalax aspalax* and the birds such as *Otis tarda*, *Syrrhaptes paradoxus*, *Aguila chrysaetos*, *Buteo hemilasius*, *Passer domesticus*, *Perdix dauuricae*, *Melanocorypha mongolica*, *Alauda arvensis* and *Emberiza fucata* are common species here.

The moderate temperate typical steppe includes Hulunbuir Grassland, Xilin Gol Grassland, Wulanchabu Grassland, the grassland of the north central of Chifeng and the Northeast Plain. The large area of grassland presented the landscape which was dominated by steppe of Tussock grass. The typical animals include *Procapra guttrosa*, *Marmota sibirica*, *Citellus dauricus*, *Ochotona daurica*, *Myospalax aspalax*, *Meles*, *Otis tarda*, *Anthropoides virgo*, *Melanocorypha mongolica*, *Aquila rapax*, *Oenanthe oenanthe*, *Calandrella cheleensis*, *Syrrhaptes paradoxus* and *Eremophila alpestris*, etc.

The desert steppe is located at the west of typical

Citellus dauricus (Photographed by Li Xiaohui)

steppe, including the west of Inner Mongolia, and some regions of Ningxia and Gansu Province. Compared with typical steppes, the heat energy is higher but the degree of wetness has decreased, the dominant components of vegetation include the Gobi Mongolian desert steppe species and the Asian central desert steppe species. Some weed, xerophytic shrub and subshrub are their companion species. The animals of desert steppe can endure drought in a long time. The *Citellus erythrogenys* of desert steppe in the north of Yinshan Mountains has replaced *Citellus dauricus pallas* of the typical steppe. The other common species include *Lagurus luteus*, *Allactaga bullata* and *Hemiechinus auritus*. *Procapra gutturosa*, *Otis tarda* and *Anthropoides virgo* of typical steppe has decreased and *Grazella subgutturosa* has increased. The *Chlamydotis undulate*, *Oenanthe deserti*, *Saxicola insignis* of birds which have the best endurance of drought appeared and they can live with little water that only comes from the food and have no other water source. *Syrrhaptes paradoxus* can sustain the lives of themselves and offsprings through fast fly and feeding. Wild camels might appear on the border between China and Mongolia in recent years. Some huge birds and animals have moved back to Helan Mountains, Wula Mountains and Kubuqi desert of Yinshan Mountains and the boundary region of China and Mongolia, such as *Procapra gutturosa*, *Capreolus capreolus*, *Meles*, *Panthra uncial*, *Felis manul*, *Felis bangalensis*, *Ovis ammon*, *Pseudois nayaur*, *Otis tarda*, *A. chrysaetos* and *Buteo hemilasius*, etc.

There are plenty of fish in the lake of moderate temperate grassland, and the *Cyprinus carpio*, *Carassius auratus*, *Leuciscus waleckii* of *Cypriniformes* are the dominant species. The cold water fish is the dominant species in the northeast zone, including *Huso dauricus*, *Acipenser schrencki*, *Hucho taimen*, *Brachymystax lenok*, *Coregonus ussuriensis*, *Esox reicherti* of *Esocidae*,

Emberiza fucata (Photographed by Yang Guisheng)

Hemiechinus auritus (Photographed by Yang Guisheng)

Eremias argus (Photographed by Xing Lianlian)

and *Lota lota* of *Gadidae*. Except for *Cyprinus carpio*, *Carassius auratus*, *Leuciscus waleckii*, many small sized fish like *Cobitidae* and *Cyprinidae* in Yellow River, *Silurus asotus* is also the common species. There are many kinds of *Amphibias* such as *Bufo gargarizans*, *B. raddei*, *Rana nigromaculata* and *R. chensinensis* in grassland and wet meadow. The *Phrynocephalus frontalis*, *P. versicolor*, *Eremias argus*, *E. multiocellata*, *Elaphe dione* and *Agkistrodon* are the dominant species of *Reptilia*.

2. Grassland Animals of Upland Meadow

The upland meadow of China is distributed in Xinjiang and the grassland vegetation takes hold in Xinjiang. Ili Grassland is one of the four alpine steppes in the world, and Bayanbulak Grassland is the second largest upland meadow of China.

The landform in Xinjiang is complex. Altai Mountains locate in the north of Xinjiang. Tianshan Mountain with an altitude of 4,000~7,435m passes through Xinjiang from the west to the east. Kunlun Mountains, Kala Kunlun Mountains and Altun Mountains are connected with Qinghai-Tibet Plateau. The Pamirs which has an average altitude of 5,000 m is located in the southwest and the Chogori with an altitude of 8,611 m also stands here. There are numerous basins and intermontane valleys among mountain chains. Under the influence of vertical climate belt and different hydrothermal condition, the vertical zonal distribution of vegetation has been formed, including temperate semi-humid grassland, typical steppe, cold temperate brush meadow and meadow steppe, sub-alpine meadow and alpine cold meadow.

Many grassland animals are distributed in various types of grasslands. Except for the common species on China's grassland, the precious huge hoofed animals such as *Equus przewalskii*, *Saiga tatarica* and *Camelus bactrianus* only distribute in the desert of Xinjiang and desert steppe zone. The *Marmota caudate* is endemic specie of Pamirs. *M. himalayana* distributes in high mountains mainly in Xinjiang and Qinghai-Tibet Plateau. *Lepus oiostolus* and *Ochotona iliensis* are only founded in alpine steppe of Xinjiang. *Syrrhaptes paradoxus* is a common species of moderate temperate grassland. *Syrrhaptes tibetanus* diffused to Kunlun-Altun Mountains of Xinjiang. *Pterocles orientalis* is only located in the northwest of Xinjiang. There are 9 kinds of *Alaudidae* in Xinjiang, including *Melanocorypha bimaculata*, *M. leucoptera* and *M. yeltoniensis*, etc., and the first three kinds of them are only found in Xinjiang, but the *Lanius minor* also can be found in the northwest of Xinjiang.

According to the difference of vertical zonality of vegetation, each species of vegetation has dominant species of animals, but some animals move among several vegetation belts.

The familiar species as follows: *Tetrax tetrax*, *Chamydotis undulate*, *Perdix dauuricae*, *S. orientalis*, *Lanius cristatus*, *L. collurio*, *Upupa epops*, *Alaudidae*, *Desert Wheatear*, *Syrrhaptes paradoxus*, *Phodopechys githagineus*, *Mustela eversmanni*, *M. altuica*, *Lagurus lagurus*, *Citellus undulates* and *Cricetulus migratorius* often appear in low upland meadow.

Lanius minor, *Aquila rapax*, *Emberiza cia*, *Petronia petronia*, *Passer domesticus*, *Passer hispaniolensis*, *Lepus capensis*, *Mustela nivalis*, *Canis lupus*, *Cuon alpinus*, *Vulpes uulpes*, *Ovis ammon*, *Capra ibex* and *Felis lynx*, etc. are located in the middle mountain cold temperate meadow steppe.

We often find *Syrrhaptes tibetanus*, *Tetraogallus tibetanus*, *T. himalayensis*, *Montifringilla blanfordi*, *M. taczanowskkii*, *M. nivalis*, *Marmota bobak*, *Lepus oiostolus* in sub-alpine meadow and alpine meadow. *Bos mutus*, *Equus kiang*, *Pantholops hodgsoni* and *Ochotona curzoniae* come from Qinghai-Tibet Plateau are located in Kunlun-Altun Mountains.

There is a famous lake wetland in Xinjiang mountain steppe and its fish resources are abundant. The *Hucho taimen* in Kanasi Lake is famous in the world. *Brachymystax lenok*, *Lota lota* and other kinds of cold water fish are not only found in the northeast of China but also in Xinjiang. Other valuable fish include *Acipenser nudiventris*, *Perca schrenki*, *Schizothorax argentatus*, *Triplophysa strauchi*, *Bufo viridis*, *Rana altaica* of *Amphibia* and *Vipera ursine*, *Agama stoliczkana* and *Phrynocephalus helioscopus* of *Reptilia*, all of which have the characteristics of grassland animals of upland meadow.

Emberiza cia (Photographed by Yang Guisheng)

Of course, the area of alpine steppe is very large and the animal community structure has difference in horizontal structure in addition to the difference of vertical zonality. Because of the differentiation of physiognomy, climate and vegetation, the composition of grassland animals is different and the fauna of upland meadow belongs to Tianshan Mountains sand sub-zone of Inner Mongolia-Xinjiang region.

3. Warm Temperate Grassland Animals

Warm-temperate grassland is located in the west of Erdos of Inner Mongolia, the north and the west of Shanxi, its altitude is about 800~1,300 m. Mu Us Desert is seated among them. The density of ravine ranks among the highest in China, forming the particular loess landform of loess plateau. The typical steppe is located in the ravine, and shrub steppe and sparse steppe are located in cleugh, hills and sands. The grassland animals of warm-temperate grassland include *Procapra guttrosa, Felis manul, Meles, Velpes velpes, Martes foina, M. flavigula, Felis chaus, Citellus dauricus, Myospalax fontanieri, Syrrhaptes paradoxus, Anthropoides virgo, Parus venustulus, Calandrella cheleensis* and *Calandrella acutirostris,* etc. The grassland fauna of warm-temperate grassland belongs to the warm-temperate desert steppe of the east grassland sub-zone of Inner Mongolia–Xinjiang region, and the southern part of Yinshan Mountains belong to the loess plateau sub-zone of North China.

4. Alpine Steppe Animals

The alpine steppe of Qinghai-Tibet Plateau appeared later than other grasslands. Qinghai-Tibet Plateau was formed by Himalayan orogeny in the late Pleistocene which is surrounded by Himalayas, Kunlun Mountains, Qilian Mountains and Hengduan Mountains, its altitude is the highest with an average altitude of 3,000~5,000 m in Tibet, Qinghai, the west of Sichuan and the southwest of Gansu Province. The landform and climate of the lofty plateau changed dramatically, with the alternate appearance of ice age and interglacial period, the Qinghai-Tibet Plateau has the alpine and dry climate with the lowest temperate of lower than -23 ℃, and the annual precipitation of only 50mm at least. With the changing of climate, the alpine steppe appeared and was divided into alpine steppe, shrub steppe, shrub meadow and desert steppe due to the different landforms.

With the appearance of grasslands, the grassland animals spread from the north and survive with biological characteristics suitable for high and cold, dry and oxygen-poor conditions. Some species evolved into some local sub-species such as *Cervus elaphus wallichi*; some species became a new species due to the changes of genetic material, shapes physical and ecological features, such as *Bos mutus*, *Equus kiang*, *Pantholops hodgsoni*, *Procapra picticaudata*, *Marmota himalayana*, *Cervus albirostris*, *Grus nigricollis* and *Carpodacus roborowskii*, etc. Moreover, the *Vulpes ferrilata*, *Lepus oiostolus*, *Ochotona curzoniae*, *Syrrhaptes tibetanus*, *Tetraogallus tibetanus*, *Melanocorypha maxima*, *Motifringilla nivalis*, *M. adamsi*, *M. blanfordi*, *M. ruficollis* and *M. taczanowskii* are mainly distributed in Qinghai-Tibet Plateau.

The adaptability of alpine steppe animals was formed under a long-term natural selection, for example, the furs with strong heat-retaining capacity are used to defend against the cold. The fibre diameter of wool of *Pantholops hodgsoni* is only 9~12 μm which is only less than 2/3 of cashmere. It is soft and stretchy. There is much air in the wool which forms effective thermal-protective coating that can maintain the heat of body. People use the light fibre to knit the valuable shawl-"shahtoosh" ("the king of cashmere" in Farsi). It is called soft gold because it is valuable. Though the Tibetan antelopes disappear as suddenly as they appear, they are often hunted by people.

The weight of *Bos mutus* could reach 1 t. The neck, body side and thorax and abdmen of wild yaks have 30~50cm black long hairs. Though the fur can not compare favourably with the fine wool of *Pantholops hodgsoni*, its black coarse wool have good marrow, which can store much air and form the thermal-protective coating, so the yak can drink ice water and sleep on snow land. The species of alpine steppe have other adaptations, for example, the proboscis of *Pantholops hodgsoni* is very wide and two sides of nasal cavity are bouffant, so the cold air will become warm through nasal cavity and then enter into lung which can protect the respiratory system. The wide grassland needs fast running of animals to escape from enemies. The speed of *Pantholops hodgsoni* is 70km/h, if they are beleaguered by wolves, they can protect themselves with sharp horns.

The fish which are able to endure the alpine environment are the dominant species of Qinghai-Tibet Plateau, such as some species of *Schizothoracinae* and *Tripophysa*. The *Gymnocypriis przewalskii* is the most famous species. The *Scutiger boilengeri*, *Nanorana pleskei* of *Amphibia*, *Agkistrodon strauchi* and *Phrgnocephalus vlangalii* of *Reptilia* are the representative species.

The alpine steppes have fostered special fauna, which is an independent Qinghai-Tibet zone in geographic zoning of fauna.

The grasslands embrace various kinds of animals for its open spirit. Meanwhile, the vast grassland also needs each animal to learn many skills of survival and breeding offsprings. When the young can not live without their parents at the difficult time, the great mother love moves every one. *Charadrius veredus* is a kind of mini wading bird which breeds in moderate-temperate grassland. They build nests in grass of ground and the young can run soon after shell smash, but they need parents to guide them to find food. One day, when we observed the grassland in a car, we saw an "injured" *Charadrius veredus* which have "broken" wings and fell down in the front of our car. We have seen through her trick, so we drove the car to her slowly, but before the car knock into her body, she flew suddenly and after several meters, she repeated the same tactics. So we stopped the car and went back and found a four-month old baby with brown fleck. This mother attracts enemies to buy time for her baby. We have moved because of mother's self-sacrifice spirit. We put the baby in our palm and gave it back to its mother. We stood far away from them for a moment, and then the babies gathered together and studied again.

Chapter 5
Grassland Eco-culture

The beautiful grassland is my homeland
Written by Huo Hua, composed by Altan-agula

The beautiful grassland is my homeland.
The wind is over blowing the green grass, everywhere are flowers.
Butterflies are flying and the birds are singing.
The blue water has caught the sunset.
Strong horses look like the color cloudy.
The cows and sheep look like pearl everywhere.
Ah ah ah, shepherd are singing loudly.
Everywhere are the happy songs.
Shepherd are singing loudly.
Everywhere are the happy songs.
The beautiful grassland is my homeland.
I love the clear river and the beautiful grass.
The grass is like the green ocean.
The yurts are like white lotuses.
The nomads are creating the happy life.
The beautiful views look like a beautiful painting.
Ah ah ah, shepherds are singing loudly.
Everywhere are the happy songs.
Shepherd are singing loudly.
Everywhere are the happy songs.

Walk into the grassland, walk into a herdsman's home

The grassland is like a big home of a herdman.

Walk into the grassland just like a herdman's home.

Dwelling Place

Stop on grassland and gaze into the distance, some white dots appear in the green leaves from the ends of the earth, when you come closer, they look like the white lotus flowers in bud. These are Mongolian yurts on grasslands, and the home of herdsman. Yurts and flowers scatter on grasslands as far as the eye can reach, brighten the grasslands, which look like a piece of magnificent and beautiful painting.

In the northern grassland region, the yurt is the unique landmark building. It was called Qionglu in ancient times. For thousands of years, the yurts have been kept out wind and rain for herdsmen. Mongolian yurt is known for its smooth round lines, unique artistic firmament-shape sculpt, and warm and comfortable room environment. The yurt is home to the northern nomads, who have travelled across mountains and rivers and migrated for generations with their footprints in the Mongolia plateau, and from ancient times to today. Although the journey was so long, the friendship established between yurts and

Mongolian Yurts on Grassland

herdsmen has never changed.

There are three reasons why the herdsmen like yurts.

1. The yurt's building materials are easy to obtain with less materials and simple workmanship

A yurt is mainly composed by five parts, Hana meshes, studdle (Wunigan in Mongolian), Taonao, the door and wool felt. First, choose the right campsite, then connect the flexible Hana meshes to a cylindrical wall, use dozens of studdles to be rafter to support the wooden and round Taonao as a skylight, and fix the bottom of studdles to the round wall composed by Hana meshes. After building the framework, cover it with one or two layers of wool felt, and set timber piles around the wool felt and firmly fasten the felt with ropes, a yurt is built.

Herdsman's Home on Grassland

Herdsmen are constructing Mongolian yurts

2. The yurt is good for heat preservation and also can prevention of wind and rain

The yurt prevents heat in summer, and defends against the cold in winter. The inside and outside of yurts seem like two different worlds all the year around. In winter, when cold wave approaches, a bitter north wind howls with a big snowfall, the yurt is sealed and warm as the stove is burning brightly; in summer, it is very hot outside, or when a fierce storm is coming, it is pleasantly cool inside the yurt because of the bad heat transition with no leakage on the top of the yurt. During the dead of night, people who live in the yurt can speak frankly opening the skylight, and lie under the skylight while watching the star and moon.

3. Convenient building, teardown and transportation of Mongolian yurt

In the past, herdsmen on grasslands have to migrate frequently all the year round, so their tiredness and painstaking are self-evident. Now they can graze on their own grasslands and still have summer-autumn campsites and winter-spring campsites. In winter-spring campsites, nomads can build houses and livestock's shelters, and they still live in yurts in summer andautumn when they graze or migrate. The important reason is that it is convenient to build, knock down and transport yurts with less time and labor.

Diet

Herdsman's dietary structure and life styles and customs are entirely different from inland rural areas.

The herdsmen like eating meat. In their daily life, beef, mutton, camel meat are staple food, but rice and flour products are complementary food. They eat three meals a day at all seasons, and often have a few drinks. Every family has self-made original tastebutter, boiled milk, dried milk cake, cheese, milk wine and fried bread stick, etc. The delicious milky tea is the necessary drink every day. When you sit in the yurt and put a spoonful of parched rice into teabowl, drinking and nibbling, you can drink it all the time every day. In spring and winter, it is all covered with ice and snow, the herdsmen get out late to graze and return home early, and they have no problem of food and clothing. But in summer and autumn, livestocks growfat, it is hot and windy, and the herdsmen have to graze early and and stay out late, so they only eat milk food and drink milk tea to fill the stomach in the daytime, and eat meat, noodle or rich after going home. Such situation goes round and round, year after year. This is the life style and customs that come form ancestors of nomads. For nomads on grassland, they are accustomed to this lifestyle and find pleasure in it.

I want to cite a saying from Confucius: "It is always good to have a friend coming from afar." Herdsman always entertain their guests by kill sheep, it is commonly seen, but what's amazing is that the technology of killing sheep, they can use the same time as killing and stewing chicken to put the mutton on the table.When drinking with a close friend, a thousand cups will be too little. They entertain their friends with good wine and delicious food, singing and dancing without sorrow and anxiety. The hospitable herdsmen believe that the life of tomorrow will be much better than today.

National Dress

National dress shows the external image and characteristics of a nation. In China, a lot of national minorities have their own national dresses with special features. The Mongol nationality, which has the largest number of population, also has its own traditional national dress. Regardless walking on the grasslands or on the streets of towns, we often see that many people wear the loose embroidered Mongolian robes with bright-colored silk belt on waist and riding boots, ladies wear upscale scarfs, and men wear hats, their warm and generous characters make people unforgettable. There are two reasons that Mongolian nomads wear national costume: One reason is that as the descendants of Genghis Khan, they have a strong sense of national pride, and the national dress is the best presentation of national image. The other is that it is closely linked with the production and living environment of grassland nomads. The climate changes frequently on grassland. In winter, northwest wind and snow follow close on another, it is extremely cold, so the thick Mongolian robe can keep warm in daytime and be used as blanket at night. In summer and autumn, the days are long and the nights short. A group of mosquitoes and flies fly in the sky, obvious insects and worms, and snakes appear and disappear mysteriously. The loose Mongolian robes and high-waisted riding boots not only keep from the burning sun, but also protect from mosquitoes and bites of snakes. Running gallop on grassland with heads held high and tightening belt can effectively avoid sick of stomach. The high-waisted riding boots can avoid injury when falling down from the horse. Therefore, Mongolian robes and high-waisted ridingboots are always the treasure of grassland nomads.

Transportation

Since ancient times, it is well-known that boats float on water and cars on road. Vehicles are important means of transportation for people living and product on land. The nomads who live on northern grasslands also have their own cars, which are different, the original cars—Lele vehicle. The Lele carriagehas been the special carriagefor nomads to migrate and transport in former days. Its wheels are big and not round, the whole structure of the carriages is very simple and it is roughly made, so it is shaky from side to side and the running speed is very slow.

Lele carriage is so different from the rest because it is closely linked with nomads' life and grasslands' natual environment. As grasslands have no forest, and only a few of trees can be found in forest steppe or mountains. Timber shortage often occurs in grasslands. Therefore, it is not easy to find suitable timber to make or repair vehicles. Furthermore, as grasslands have no roads and are a vast territory with a sparse population, so nomads do not care about whether the wheels are round or not, but they pay more attention on whether the vehicles can pass through hilly land, sand or muddy wetland. The grassland is home to nomads, and they cherish every tree and bush. The Lele carriage will not leave track on grasslands, and can protect the grassland vegetation from being destroyed.

In 1950—1960s, on the grassland of Inner Mongolia, we often saw 10~20 Lele carriages loading animal by-products or articles of daily life slowly cross the streets of some towns. When you take a closer look, they look like simple large-wheel wooden vehicles pulled by cows, but when looking out, they just like trackless trains with goods. This type of train only has starting and terminal stations without schedule and people on the train do not need to worry about time and behind schedule, because it has no running time. The tail of carriages is hung with metal buckets and peelers, which collide with each other and send out the noise of "ding-dong", and the drivers in the front are relieved and hurry on their journey in the nighttime without looking back.

Although the Lele carriages run slowly, these big wheels are keeping running, and help nomads through many years from ancient times to now. Nowadays, Lele carriages are replaced by convenient and fast modern vehicles such as tractors and cars. The Lele carriage is

Vehicle of Herdsmen-Lele Carriage

coming to museum of history step by step. Without the regret, it has shortened the hope-road that Mongolian herdsmen have pursued from generation to generation and have brought them a lot of happy memories.

Unique Charm of Grassland

Dating at the Mongolian Festival

The full moon on the 15th day of the month has risen to the sky.

But why is there no cloud next to it?

I'm waiting for my beautiful girl.

Why have you not arrived here yet?

If it doesn't rain from the sky.

The crab apple tree will not blossom by itself.

If only I, your lover, wait patiently,

The girl in my heart will then come to me.

Legend of Aobao

When you drive on the grassland of Inner Mongolia for the first time, you will feel relaxed and happy, with higher sky and vast land, everything in your eyes is fresh. Wherever you go, you may often see some stone piles which are piled up by stones. This is Aobao in Mongolian, which means "pile" in Chinese.

In the book "Seek Dreams in Grassland", Gai Shanlin conducts a detailed study on Aobao.

Herdsmen usually construct Aobao on the high lands, mountain passes or some places where they are easy seen by people. Aobao can be constructed individually, or combined by seven Aobaos—the bigger one in the centre as the main body, and three small ones on both sides; or combined by thirteen Aobaos. The greatest size of Aobao group is in the south of an old town named Alunsumu of Damao County, which is made of 36 Aobaos in-line. The shape of most Aobaos is circular, but their height and size are different. Some Aobaos are several meters in height with willow twigs on the top and prayer flags in Sanskrit. They look like spires from far away, which are quite spectacular with a sense of mystery.

In the mind of the Mongolian, Aobao is sacred, spiritual and revered by the herdsmen. Every time the herdsmen go far away, when they pass through Aobao, they would worship Aobao and add some stones to Aobao for the purpose of praying for the pasturing

Dating at Aobao

area, hoping that the gods of the land and mountains and rivers will bless them.

There are three reasons why Mongolian revered Aobao.

1. Raising livestock depending on nature

The Mongolian nationality calls them "the nation on horseback". They lead the nomadic lives from generation to generation on the vast grassland, move frequently throughout the year without definite residence, this may be called "take the sky as quilt, the earth as bed, and where there is water and grass, there is home." The most important means of production are horses, cows, sheep and camels, and other domestic animals. If it's a good year andit has favorable weather for crops, and number of livestock will be increased, and nomads will enjoy their lives. However, severe natural disasters occur, such as drought, snowstorm, pest disaster and some kind of infection between people and livestock, and lack of necessary natural science knowledge make the nomads submit to the will of heaven and manipulated by the nature as the transportation is inconvenient.

2. Living depends on the nature

In the past, the means of livelihood of herdsmen in grassland region such as foods, clothing, residence, transportation, and needbasically come from the livestock and livestock products. With the exception of a small amount of grain, salt and tea, a large amount of means of livelihood can be solved by herdsmen. As for the poor and the rich, it cannot be governed by herdsmen, because they are unable to fight against nature, but only passively wait for the mercy of nature. Thus, it is natural that people take personal fate connected with the god, believe in the philosophy of "life and death are decreed by fate, wealth and rank are matters of destiny", and pin their hopes of future on gods.

3. Worship Many Gods

Mongolians believe that there are different gods of the heaven, the earth, the mountains, plains, rivers and other places. Since people have their residence, there should have the places for various gods to live. In the grassland region, since it is a scarcely populated area, and the gods exist everywhere at any time, the best and most feasible way is to let fortune teller choose the best place to build Aobaos for various gods. Therefore, on the Inner Mongolian grassland and in the minds of the Mongolian, the mystery of Aobao is that it is the carrier and expression of gods.

In addition to the above-mentioned legends of color of religious superstition, in all ages, in the herdsmen's production and life, Aobao also has many important functions, such as boundary markers, distance markers, the resident markers, road and direction markers and so on, meanwhile, Aobao is also endowed with full-bodied national culture.

Since ancient times, in Lunar May or June of every year, people in various places in grassland hold Aobao feast for the purpose of giving thanks to Aobao for bringing them bright and gifts, praying for both the growth of population and stock breeding thrived, and celebrating a good harvest in sight. At this time of grassland, the cold winter and spring have gone, sunlight warms the land, forages floyrish, birds twitter and fragrance of flowers, livestocks grow stout and strong, and everything is fresh again. Herdsmen from all quarters come to attend Aobao Festival spontaneously, riding tall horses, wearing the best clothes, taking along good food, brick tea and Mongolian chess and other entertainment stuff, to eat, drink, sing and dance together and enjoy the festival.

The content of Aobao Feast contains two main elements: worship ceremony and athletics and performance.

There are four forms of sacrificial rites of Ancient Aobao: Jade sacrifice, fire sacrifice, blood sacrifice and wine sacrifice. Jade sacrifice is to bury exquisite jade and expensive jewelries under Aobao; fire sacrifice is to put all kinds of food into a fire in front of Aobao; blood sacrifice is to slaughter of a sheep and cow in order to put them in front of Aobao; wine sacrifice is to sprinkle milk wine, and white spirit in front of Aobao. Nowadays, blood sacrifice and wine sacrifice are still popular. In worship of Aobao, it usually have lamas chanting, pastoralists bowing on bended knees to show there pieties. Athletics and performance mainly contains horse race, singing and dancing, etc.

From the ancient time, Aobao has made significant influence on the production, life and ideology of normads on the grasslands. With development of society, and the popularization of scientific and cultural knowledge, the mystery of Aobao has been gradually weakened, at present, Aobao and Aobao Feast have become the place for local people to celebrate a bumper harvest, promote friendship, and communicate with each other.

Caballero of Grassland, Wulanmuqi

Wulanmuqi, translated into red literary art team in Chinese, is a model to spread national culture and art in China's minority area. They are praised as "the art light cavalry on the grassland" by prairie people.

The Inner Mongolia Autonomous Region covers a vast geographical area, and is sparsely populated with inconvenient transportation. In some pastoral areas, people can meet one or two Mongolian yurts every other dozens of miles. The herdsmen's cultural life is very poor, they want to communicate with the outside world and know the outside world. In order to satisfy the spiritual needs of extensive herdsmen, the first Wulanmuqi came into being in 1957 in Sonid Right Banner of Xilin Gol League grassland. At that time, Wulanmuqi had 12 members, all played instruments: one piece of accordion, Sihu, Matouqin (horsehead shaped string instrument) and Mongolian flute. They go deep into pastoral areas and Mongolian yurts on carriages with these simple instruments, books, radios, epidiascopes, pictures and outfits. Wulanmuqi members perform fervently various programs of entertainment for herdsmen, teaching the knowledge of current affairs and science, teaching herdsmen's children singing and dancing, lending books, cutting hair and delivering medicines, making friends with the herdsmen, thus they were well received by nomads. The literary form and activities of Wulanmuqi have lasted for 51 years, and eternally renewed.

The Wulanmuqi performing team is terse and forceful, consisting of more than ten people, and the member comes from the local herdsmen. The Mongolian nationality is skilled in singing and dancing and endowed with many gifts. The Mongolians are natural singers and dancers. Everyone member of Mulanmuqi masters many skills while specializing in one, they can not can sing, dance, tell story, but also play many musical instruments, and have the skills of editing, di-

Epigraph of Guo Moruo for Wulanmuqi

Wulanmuqi actors are singing

Wulanmuqi actors are performing

recting and acting, so they excel at almost everything. Most material of Wulanmuqi comes from grasslands and pastoral areas, which show Mongolian national customs and regional characteristics of Inner Mongolia as well as the production and life of postoral areas.

The performance of Wulanmuqi is regardless of harsh winter and hot summer. They take blue sky as stage curtain, and grassland as stage. They will perform as scheduled so long as asked by herdsmen and no matter how many audiences will appreciate. Their programs will not be repeated even performing for a few days. "The Sun Rises on the Prairie", "Chairman Mao, we prairie people love you", "Song of Praise", "Dating at the Mongolian Festival", "The beautiful grassland is my home", "Love of Grassland", "Beautiful night of grassland", "The Mongolian", "Engraved Saddle" and "Toasting Song", etc. make people never get tired of hearing. Some graceful national dances are never get tired of seeing, such as "Mongolian Bowl Dance", "Chopsticks Dance", "A Milkmaid", "Ladies on Grassland", "Galoping Horses", and "Andai Dance". They praise the beautiful and magnificent grassland, the hardworking of grassland people and long for future happy life by singing loudly and dancing gracefully. Herdsmen seem to see their familiar grasslands and homes through the performance, they also see the silhouette of themselves. The sight stirs up their feelings, and makes them feel so happy.

Premier Zhou Enlai once said that we should ignite Wulanmuqi all across the country. Under his leadership, Wulanmuqi was just like a running horse,

Fair-sounding Horse-headed string instrument

went beyond the grassland for the first time, and then headed for the whole country, and every corner of the world.

Songs of Wulanmuqi were heard the whole world. Wulanmuqi has become the pacesetter of spreading national culture and art, and got the reputation of "grassland light-horseman".

Grassland Fair—Nadam Fair

With regard to Nadam Fair, nobody living in the grassland doesn't know it. In Mongolian, Nadam means game or entertainment, which is the annual traditional grassland fairfeast of the Mongolian nation. In July and August of each year, in the grasslands of Inner Mongolia and Xinjiang and other places where inhabited by the Mongolian nation, Nadam will be held. It is the best time of the year on grasslands: The sun is bright and high in the sky, it is fine and warm, and the grass looks like a green carpet. The green and boundless grasslands look like a boundless stretch of green sea, crowds of birds fly past the sky of grasslands sending out sweet sounds. Accompanied by melodious singing of herdsmen, the stout and strong cattle and sheep play with each other, or graze quietly. At this moment, the herdsmen enjoy themselves and refreshed, they wear neat and beautiful festive costumes, bring along with the yound and the old, ride on horses or drive cars, bring food and beverages, regardless of the long distance of the journey, rush to Nadam Fair from all quarters. The quiet grasslands in ordinary days suddenly boil up with colorful flags fluttered everywhere. Songs, cheers and noises sink into a sea of joy.

As the traditional festive of Mongolian nation, Nadam fair has a long history. Before the 12th century, Mongolian leaders held meetings in addition to formulating laws and regulations, appointing and dismissing officials and punishing and rewarding, they held large-scale Nadam events. Wrestling, archery and horse racing had become the main activities at that

Nadam Fair

time. Mongolia people like to participate in and watch traditional wrestling, archery and horse racing and other athletic competitions in Nadam fair, which are closely related to the ways of production and their lifestyles inherited from their ancestors.

From the ancient time, the nomads on the Mongolian Plateau made a living by hunting and or raising animals. During the long-term fighting with a variety of animals, they gradually learned and mastered superb wrestling skills. It can be said that wrestling is one of basic skills of nomads in beating ferocious beast. The author has more than once witnessed the thrilling scenes in which steppe nomads fought with adult cattle in bare-hand, and made cattle heavily fall on the ground. After humen invented bows and arrows, nomads living in grasslands took horse riding and archery as an advanced way of hunting which was widely used in grasslands. Because horse riding and archery are more secure for pastoralists to fight with beast comparing with wrestling and fighting in bare hands, and have a better result in hunting. Horse racing is the strength of the Mongolian people's athletic contest. Mongolian people call themselves the nation on horseback. From the ancient time, regardless of men and women, old and young, everybody is excellent rider, and horseback hero. Riding horses on grassland is just like city people riding

Horse Racing on Grassland

bicycles in the city. Because the nomads start on horse riding training from childhood. Parents take children to horseback personally, and let children walk alone on horseback, riding alone or riding with their parents to graze in grassland, soon after, the bold children are able to gallop on the horseback with or without saddles, becoming qualified horsemen.

With the development of society, the nomads on the Mongolian Plateau not only use the unique skills like wrestling, archery and horse racing as an necessary means of hunting to make a living, but also gradually expand their social functions to the military training of the territory exploration and homeland defense, make them become important subjects of military training of the Mongolian nation. The government of the Yuan Dynasty (1271-1368) set up a special institution called "Xiao Shu" to manage the wrestling business nationwide. In 1279, the Yuan dynasty perished the Southern Song Dynasty (1127-1279) and unified China, since then, Nadam feast was carried out widely on the Mongolian Plateau, and wrestling, archery and horse racing were the indispensable competition events. Since the Qing Dynasty, Mongolian Nadam fair has become the official organized large entertainment activity, and the activities, content and scale have changed a lot. In the 25th year of emperor Kangxi (1686) of the Qing Dynasty, Emperor Kangxi called up nobilities of 10 banners such as Hexigten Banner, Baarin Right Banner, Baarin Left Banner, Alute Banner to hold large-scale Nadam fair, which was well-known "Zhaowudatala Gala" in history. In 1771, Wobaxi led Torgud Mongolian tribes, to go back to the motherland from the Volga River after experiencing numerous difficulties and dangers. In September in Chengde Imperial Summer Resort, Emperor Qianlong called up leaders of all ethnic groups, the nobilities of all leagues of Inner Mongolia to hold unprecedented Nadam fair. The main contents include the wrestling, archery, horse racing, and other cultural and sports performance and hunting activities. Emperor Qianlong attended the Fair in person, and awarded his ministers. Mongolian nobilities of the Qing Dynasty held Nadam Fair every six months, one year or three years, and winners were rewarded and titled. After new China was founded, the traditional wrestling, archery and horse racing in Nadam fair have gradually lost the former functions, but highlighted the national spirits of courage and bravery of Mongolian people. Now, Nadam Fair is not what it used to be. More and more activities and the larger scales make it become an extravaganza of Mongolian grassland culture, sports, entertainment and arts. In addition to the traditional wrestling, archery and horse racing, equestrianism, riflery, judo, motorcycle racing and Mongolian chess match are added.

On the playing field of annual Nadam Fair, the material exchange activities have never been more active around the playing field. All kinds of daily necessities, consumer goods, food and beverages are sold restaurants, tea stalls, bookstalls, story-telling tents, and other services and entertainment activities gather around the site. Wulanmuqi performing team,

Mongolian style wrestling

film projection team and technology service team also come to add to the fun. Nadam Fair is filled with people, gathering together to celebrate the yearly festival. Nadam Fair is held once a year, but the content is increasingly diverse and splendid year by year.

Advance with the times, march toward modernization

Like magic horses with wings, we gallop on the grassland.

Aha hey,

Grassland is so green, water is so clear and sheeps and cattles are strong Goodbye, my green grassland, goodbye, my beautiful hometown.

Aha hey,

We fly to distant places like swallows for great ideals.

Like magic horses with wings, we gallop on the grassland.

Aha hey,

Chimneys stand high in the sky, flowers enclose factories.

Goodbye, my green grassland, goodbye, my beautiful hometown.

Aha hey,

We will fly back to Baotou steel to build it.

Like magic horses with wings, we gallop on the

grassland.

Aha hey,

It will be no longer wild but steel manufacturing.

Goodbye, my golden grassland, goodbye, my happy hometown.

Aha hey,

We are to be iron workers and devote my young to Baotou steel.

Time passes very quickly. The outlook of grassland and lifestyles are going through irreversible fundamental changes.

Grassland Law of the People's Republic of China was adopted on June 18, 1985; revised on December 28, 2002 and implemented by the State Council on March 1, 2003. It precisely states that the right of use of meadow belongs to farm household. Since then, the Mongolian herdsmen end their nomadiclife of living where there is water and grass inherited from their forefathers and start a quiet and easeful life.

Herdsmen ride on horses and whip on horses to advance, followed by Lele carriage loading people and cargos, migrate to change residence in long journey ("Zou'aote" in Mongolian), which were commonly seen in the past in grasslands have gone for ever. This has been replaced by many settlements and new villages with different scales. The settlements are made up with many herdsmen family. The masonry-timber structure houses and sheds are cool in summer and warm in winter, and durable in use. Now, overhead power lines, wind-driven generators are available. Power networks are accessible to towns and settlements of grasslands. The electricity has provided strong impetus foreconomic development of grasslands, especially the light for grassland people. The modern facilities enjoyed by city people in the past are now accessible to herdsmen, for example, electric lights, telephones TVs, radios, VCDs, mobile phones, washing machines, refrigerators, and computer, etc. Now, even ordinary herdsmen have all modern articles like city people. If you carefully note the changes occurred on the grasslands, you will undoubtedly believe that settle-down is a historic turning point in grassland pastoral areas, and it has a profound and long-lasting impact on the grasslands and the people.

Certifications of use right of grassland gave herdsmen an incentive to protect and rationally use grasslands. Because the free grazing is limit by laws, the production activity of herdsmen only can be carried out in their own pasture. Therefore, the animal husbandry production mode of grazing in cage, semi-drylot feeding and drylot feeding are common in grassland pastoral areas. As the production mode has been changed, herdsmen must open up new source of fodder to increase the yield of fodder grass. It has become the conscious action to grow grass and forage artificially on a large scale. In order to raise economic benefit, they have to improve the quality of breed, raise fine breed, shorten the growing stage, and increase the crop rate. The idea of scientific feeding and management has got popular support.

From nomadism to settlement, national education has the greatest and far-reaching influence on grassland herdsmen. Primary and secondary schools were built in settlements and new villages of herdsmen, making the national education of prairie pastoral areas popularized, and more children of hersmen get access to formal education. Beyond all doubt, this move has laid a solid foundation for improving the cultural quality of the Mongolian nationality and training specialist of national minorities. The horseback school that had long enjoyed a good reputation quietly disappeared.

From nomadism to settlement, another obvious change is the increase of population and improvement of health level. Everywhere of grasslands shows the scenes of serenity and prosperity. During the nomadic stage, as grassland environment was difficult, living conditions were poor, living standard was low, medical treatment and public health had no guarantee, the birth rate was low and the death rate was high and natural growth rate of the population was slow. After settlement, health centers or clinics had been set up in bigger settlements and new villages. The common diseases could be treated in health centers. The herdsmen's health conditions have been improved.

Chapter 6
Grassland Eco-tourism

The beautiful night scene of grassland
Written by Bai Jie, composed by Wang Hesheng

The night scene of grassland is beautiful,
Instrumental is melodious, tune of flute cheerful.
Evening breeze blows the stars of the Milky Way,
My felt yurt is full of their brilliant silver ray.
Ah ha ho eh ah ha ho,
Evening breeze blows the stars of the Milky Way,
My felt yurt is full of their brilliant silver ray.
The night scene of grassland is beautiful,
High above the heaven is rising the moon delightful,
Evening breeze tenderly caresses my green dream,
Cattle and sheep roam the border as cloud agleam.
Ah ha ho eh ah ha ho,
Evening breeze tenderly caresses my green dream,
Cattle and sheep roam the border as cloud agleam.
The night scene of grassland is beautiful,
Before raising goblets we're already blissful.
Evening breeze is singing a sweet song,
The moon is so charming that cavaliers linger long.
Ah ha ho eh ah ha ho,
Ah ha ho eh ah ha ho!
Evening breeze is singing a sweet song,
The moon is so charming that cavaliers linger long.

Welcome to grassland

Grassland of China is mainly distributed in the temperate zone of the Northern Hemisphere. Grassland of China looks like a wonderful poem, expressed in perfect harmony, which makes the people praise it greatly. It also looks like a gorgeous and colorful painting, which is amazing and brilliant. China's grasslands extend from Songnen Plain of northeast, across the Inner Mongolia Plateau and Loess Plateau to the western Altai Mountains from east to west. The grasslands stretch for thousands of miles, standing on the northern edge of China's territory. In addition, grassland looks like a green carpet hanging on the southwest of China.

All guests from afar come to the grassland with a pure childlike innocence. They long for knowing the past and present of grasslands of China. However, the grasslands of China are too big to view the whole scenery, peculiar national customs and relics in a short time. In this chapter, we select some typical and representative interesting news to amuse readers.

Hulunbuir Grassland

A forest of Xinganling.
Where the brave Oroqen lives here.
With one horse and one gun.
They hunt countless roe deers.

Hulunbuir Grassland is located at the east of Inner Mongolia Autonomous Region, with the coutinu-

Daily life of Oroqen herdsmen

Daben Lake

Horse traction and sledge

Huhe Nur Steppe

ous Xinganling leaning close to forest steppe, meadow steppe and upland meadow with plenty of water and lush grass; it looks like a huge dragon shaking heaven and earth.

The name of "Hulunbuir" comes from the names of two lakes, Hunlun Lake (also known as Dalai Nuur since Qing Dynasty) and Buir Lake. "Hulun" meas *Lutra lutra* in Mongolian, "Buir" means male *Lutra lutra*, the two lakes are famous for teeming with *Lutra lutra*.

Accroding to the book *History of Inner Monglia* written by the famous historian Jian Bozan, Hulunbuir Grassland is the cradle of minorities of China. In China's history, the minorities of Xianbei, Qidan, Nuzhen and Mongolian grew up in this cradle, spent their childhood in Hulunbuir grassland. They opened the gate of the Great Wall just from here, and then cross the Yellow River, and went on the historical stage. And now, the minorities live in Hunlun Buir Grassland mainly include Mongolian, Oroqen, Ewenk and Daur.

The four seasons in Hulunbuir Grassland are not clear, with long and cold winter and short and warm summer, or sometimes even no summer. In winter and spring, all things in Hulunbuir grassland are clad in silvery white. In the natural ski fields, guests can ski or ride sleds driven by horses across the forest and frozen grassland. In summer and autumn, mountians and plains in the grassland turn green, and colorful flowers and grasses are scattered everywhere. The garden burnets sway from side to side in breeze. *Sanguisorba officinalis*, *Ixora chinensis*, *Paenoia albiflora*, *Consolida ajacis*, *Dianthus chinensis*, *Trollius chinensis* and *Anemone cathayensis* contend in beauty and fascination in front of visitors; *Iris ensata*, *Hemerocallis citrina* and *Filifolium sibiricum* costmary look like brocade knitted by fairy. Crowds of birds fly over the grassland and woods, and horses, cattle and sheep eat grass on the shiny green grassland peacefully. Several rivers united by spring and streams flow on the grassland tramping over hill and dale. Fair-sounding long-tune Mongolian folk songs resound on the grassland and forest.

If you have any chance to tour on Hulunbuir Grassland, the first place you should visit is Huh Nuur which is 61km away from Hailar City. It is a typical example of scenery of Hulunbuir grassland, where guests can experience the production, life and customs of minorities. You can wear national costumes, ride horses and camels, ride Lele carriages on the grassland, fish in the lake by yacht, and hunt in the nearby forest. When you have a rest, the hosts may provide you with traditional cate, i.e. finger mutton, roast leg of lamb, instant-boiled mutton, milk tea, fermented milk, and all kinds of milk food. Beyond that, another place to see is Hunlun lake, which is the largest freshwater lake in Inner Mongolia and the fifth largest lake in China. Hunlun Lake lies among New Barga Left Banner, New Barga Right Banner and Manzhouli City with the length of 100km and the width of 50km and perimeter of 400km. It has the total pondage of 13,000,000,000 m^3. Data show that Hunlun Lake has 240 kinds of birds, accounting for one fifth of national total. It may be called the heaven of birds. The surrounding region of the lake is a rish pasture. Guests may not only have a good taste of delicious fish, but also can see kinds of birds, beasts and wild flowers. Your journey to Hulunbuir will leave you with good memories.

Xilin Gol Grassland

Inner Mongolia grassland is the main body of temperate grassland of China. Xilin Gol grassland is a typical example of Inner Mongolia grassland. The reason is that Xilin Gol grassland is composed of typical

Xilin River

Xilin River

Four seasons of Xilin Gol Grassland are clear and majestic. The best seasons in one year going sightseeing on grasslands are summer and autumn. No matter where the tourists are from to step on Xilin Gol grassland, the first impression is high sky and broad land, like walking into the heaven of herdsmen, with fresh air, fragrant pasture spraying into one's face, refreshing and relaxed. Looking around the surroundings, green mountains and rivers are embroidered on the vast grasslands. Looking into the distance from a high place, the boundless grassland lies beneath the boundless skies. Scenery of the grasslands which has been tracking down a thousand times in the dream is right in front of you: in the blue sky float the white clouds, under which the horses gallop. Crack my whip, sending the sound to everywhere, and flocks of birds begin to fly up.

steppe, meadow steppe, desert steppe and sand vegetation. Moreover, Xilin Gol grassland also has many historic relics from the ancient times. The original Mongolian lifestyle and customs have been preserved. Xilin Gol Grassland Natural Protection Area was classified as international biosphere by United Nations Educational, Scientific and Cultural Organization (UNESCO).

"Xilin Gol" means a river lies in hilly land in Mongolian. Most of the rivers of Xilin Gol are interior rivers, including Wulagai River, Balagen River, Xilin Gol River and Gaogesitai River. The outflow rivers include Luanhe river system. There are more than 1,300 big or small lakes, and the total pondage within the territory is 3,500,000,000m^3. These rivers and lakes flow quietly on grasslands year in and year out, moistening the vast grasslands.

Xilin Gol Typical Steppe (Photographed by Xu Zhu)

Herdsmen on Xilin Gol Grassland

Ceremony of Roast Whole Lamb

Xilin Gol Grassland has more than 20 nations with the total population of more than 900,000, including Mongol nationality, Han, Hui, Daur, and Oroqen. About 1/3 of them is Mongol nationality. Mongolian people is composed of six Mongol tribes in history, namely, Ujimqin tribe, Haojite tribe, Abahanar tribe, Abaga tribe, Snid tribe and Chahar tribe, Among them, the Chahar tribe is the tribe where Khan of Mongolia's lived since Genghis Khan encamped here. The Abaga tribe is the descendant and subjects of Belgutei who was the younger brother of the Genghis Khan. Nowadays, the modern lifestyle has already seeped into each corner of grasslands. However, from several aspects such as clothes and ornaments, food and drink, residence, marriage, etiquette, festivals, songs and dances, you can still see the nomadic culture and national customs of grasslands with a long history.

There are many historic relics and natural landscapes on Xilin Gol Grassland, including ancient Great Wall sites of Qin, Yan and Jin dynasties, Capital Site of Yuan Dynasty, Beizi Temple which is one of Inner Mongolian four big temples, and Tunggur basin which is famous as the hometown of dinosaur. Hui-Teng-Liang (meaning cold in Mongolian), which lies in the southern part of Xilinhot, is the idealist place to appreciate the scenery of grasslands. Tunggur basin near the Erenhot City is famous as the "animal fossil treasure-house" in the world. From the middle of 17th century to the middle of the 20th century, invaders plundered a large quantity of the dinosaur fossils here in the name of inspection, and now they are stored in many museums around the world.

Good Scenery of Ili Grassland

The Xinjiang Uygur Autonomous Region lies in the western temperate desert area of China. Ili Grassland is the largest natural oasis in the north and south of Xinjiang Tianshan Mountains. It boasts the special geographical location, topography and landform, and excellent ecological environment, which forms sharp contrast with the surrounding desert. The leisure and recreational environment, unique natural landscape and ethnic customs attract a large number of Chinese and foreign friends and visitors each year for sightseeing, inspection or holidays.

In arid land ecosystem, the barren desert, just the same as the forest and grassland, occupy an influential organism's habits place. Its distribution, growth and decline follow the certain law of nature, which is the fact and unchanged with the will of human. In the desert region of Xinjiang with less precipitation and high evaporation, the appearance of Ili Grassland, such a natural oasis full of vigourIli makes many people beyond belief. Anyway, the existence of Ili Grassland, Iliand its gorgeous natural scenery are the outcome of nature, which is associated with its geographic location, natural environment and peculiar microclimate. Ili Grassland is located at the foot of the western Tianshan Mountains in Xinjiang, bordering on Kazakhstan in the west, adjacent to the Junggar Basin in the north and Tarim Basin in the south, and extending east-west across Tianshan Mountains. Ili Grassland looks like the huge body of a butterfly, which encircled in the center by two big Basins like the butterfly wings. Tianshan Mountains have an altitude of about 3,000~4,000m, but the altitude of Ili River valley bottom is less than 1,000m, which is the truly alpine and gorge region. Mountain spring covers densely on two sides of Ili River. The gurgling streams, melting snow of high mountains make the swift the current Ili River flow at great speed all the year round. Inside and outside of Ili river valley are two different worlds. Under

Ili Grassland

Desert Steppe

this natural pattern, there is nothing to be surprised that Ili Grassland appears in barren desert area like lotus out of water.

The Ili Grassland includes upland meadow and river valley grassland around the Ili River. Nalat Grassland, Tangbula grassland, Kangnaiz Grassland and Zhaosu Grassland have higher awareness. Every piece of grassland has distinct characteristics or touching legend.

1. Tangbula Grassland

Tangbula grassland is located at Kashgar gorge within the territory of Nilka County of Xinjiang. "Tangbula" means "seal" in Kazakh. On the east edge of canyon, there is a piece of very large rock, whose form is exactly like a privy seal, hence the name. Tangbula is a natural scenic spot composed of forest, grassland, hillstone and streams. It is said that there are 113 ditches, 113 sceneries on the Tangbula Mountain, which are scenic and infinite. Kashgar River, running east and west, the two sides of north and south are high mountain ridges, imposing, grand sight, tilted footpath streams, natural hot spring, high mountains, lake, fantastic rock peaks, and extremely skillful stone gates and bridges can be found everywhere. Petroglyphs and Wusun ancient tombs in Tangbula were discovered by the archaeologists. Now it has been developed as the famous scenic spot and summer resort in Xinjiang. The tourists can not only have a taste of natural power and the charm of Tangbula grassland, but also track the footprints made by the ancestors in the ancient times.

2. Kangnaiz Grassland

Kangnaiz Grassland lies in the Xinyuan county of Xinjiang. "Kangnaiz" means "sun slope" in Mongolian. There are dense forests, rivers converged from streams, grasslands with clear water and lush pasture, wild apple forest with the largest area in Asia and Europe region handed down from the Middle Ages. Many valuable and rare wild animals such as snow leopard, silver fox, snow cocks, red deer and other rare wild animals also haunt about Kangnaiz Grassland. Kangnaiz Grassland is the four-season pasture with both upland meadow and river valley grassland. The scenery is enchanting in all seasons, especially in the spring. The famous Xinjiang fine-wool sheep is born here. The Qiahepu waterfall, which is about 3km from Xinyuan County, is one of the most important natural sights in Kangnaiz Grassland.

3. Zhaosu Grassland

Zhaosu Grassland lies in Zhaosu County of the western Xinjiang, which borders with Kazakhstan. "Zhaosu" in ancient Chinese means reviving and come to life, and universe in Mongolian, named after a largest Tibetan Buddhist in Xinjiang-Shengyou Temple. Zhaosu Grassland is located in an intermountain basin of the Central Asian steppe with the altitude of 1,323~6,995m. It belongs to the continental temperate mountainous semi-arid and sub-humid climate. Winter is long and cold, and there is not summer.

In the history, the ancient indigenous people, the Serbs, Dayuezhi, Wusun, Turkic, modern Mongolian and Kazak all had or have lived on this piece of grassland. It is the hometown of "Wusun" in China's Han Dynasty. Colorful multi-ethnic culture mingled here in different times. This place is one of the birthplaces of the Xinjiang grassland culture. Tengger Peak, "the Father of the Tianshan Mountains", lies on the border between China and Kazakhstan with an altitude of 6,995m, and the depth of snow about 100 km^2 all the year round. It is one of the important water sources of Tekes River, which is the largest branch of the Ili River. Zhaosu Grassland is the main part of the ancient Silk Road. It is also the goods distributing center in the ancient Western Regions. Zhaosu is world-renowned for producing Tianma, which has been praised as "Tianma's hometown."

4. Nalat Grassland

Nalat grassland is located in the east of Xinyuan County in Xinjiang. It links Xinyuan County and Yining City in the west, and then goes straight to 312 National Road. It connects Baluntai highway in the east. The Kuqa highway is also connected with Nalat Grassland.

"Nalat" means "the first place you can see the sun" in Mongolian. When Genghis Khan conquered the western China with the subordinates, one troop of the Mongol nationality cut through Tianshan Mountains from south to the north heading for Yili City. At that time it was the early spring, the mountains were covered with snow, officers and men were tired and hungry. Unexpectedly, the green grassland with sunshine is in sight after climbing over a mountain, they exclaimed "Nalat" (a place where you can see the sun), hence this grassland is named "Nalat".

According to the research, Kazak people had begun to make a living by herding on Nalat grassland since the Western Han Dynasty. Due to the large population Kazak, the Nalat Grassland is also called "The cradle of Kazak".

The Nalat grassland is one of the world famous high mountain river valley grasslands. High mountains and river valleys interwave vertically and horizontally. The natural view is varied in Nalat Grassland. The Nalat Grassland is the sub-mountain meadow vegetation developed on the accumulation layer of the Tertiary period. Its annual amount of precipitation is 800 mm.

Nalat Grassland

Children of Kazak Nationality living on Nalat Grassland

Forest vegetation and grassland vegetation grow well with a variety of plant species, which provide an ideal ecological environment for wild animals. In history, the Nalat grassland was also called "deer park", has been one of the important pastures in summer. From the beginning of second month of spring every year, wild grasses and flowers such as *Iris tenuifolia*, *Phlomis umbrosa*, *Gentianella Moench*, *Thymus quinquec ostatus*, *Fragaria ananassa*, *Agropyron cristatum*, *Festuca ovina* and *Carex tristachya* begin to break through the soil and burst into bloom, bring tender green for the grassland. During the winter, when the high mountains are covered by snow, you can view the snow-covered landscape on the grassland, skiing, riding horses or the snowfield motorcycle across the grassland. You can experience a specific scene of snow in the river valley. Now, The Nalat Grassland has been ranked as "3A" national tourism scenic area, which covers an area of about 800 km².

Statue of Wang Zhaojun

Zhaojun Tomb Written by Dong Biwu

Relics on the grassland

Since ancient times, grasslands in northern China are the cradle of northern nomadic people, the stage for creating the history, and the homes built by multi-ethnic nations. In the long history, the ancestors of the Chinese nation created many colorful grassland cultures and chapters of history with their life and blood on the vast grassland. The dust-laden historical traces are found on grasslands. Due to the limit of space, this chaper has only a limited view of the history of grasslands.

The ambassador of peace——Wang Zhaojun

Zhaojun Tomb is a must-see sight spot for Chinese and foreign visitors who travel to Inner Mongolian Grassland or Huhhot. As a symbol of national unity, Wang Zhaojun has got a reputation for a long time. The history of the tomb is over 2,000 years. As one of the eight popular tour destinations in Huhhot, the tomb is a famous cultural and historical scene in Huhhot.

Wang Zhaojun, named Qiang, is commonly known by her style name Zhaojun. She was born into a prominent family of Zigui County (now Xingshan County, Hubei Province), a maid of the palace of the Western Han Dynasty. In 33 B.C., Huhaanyeke took the opportunity to ask the emperor to be allowed to become an imperial son-in-law. Wang ZhaoJun volunteered to marry Huhaanyeke Khan and she was made Ninghu Eshi (wife of Khan). Because of this marriage, battles ceased between the Huns and Han Dynasty for 50 years. This move encourages the understanding between two different cultures and has positive effect on both social and civil aspects. Therefore, this legend has been passed down from generation to generation for over 2000 years. Her sacrifice and braveness still moved people in nowadays.

Zhaojun Tomb is located at southern suburbs of Huhhot City. The tomb is about 33m high. Surrounded by green pines and weeping willows, the surface of the tomb is extremely striking. In late autumn when grass and trees wither, those plants on the cemetery mound continue to prosper, hence it is called the "Green Mound".

As soon as you walk into the Zhaojun park, bronze statues of Wang Zhaojun and her husband (Khan) will jump into the sight, depicting a vivid picture of them talking lively and jovially. The statues are the symbol of the friendly relations between the Han nationality and the Hun nationality, which was greatly promoted by Wang Zhaojun. Behind the statues, there are several steles displaying the praises marked with a poem of Dong Biwu, which highly praises the historic feat of Wang Zhaojun. Behind the steles, stone stairs and a stone platform are connected with the Zhaojun Tomb. The two sides of the tomb are historical relics, exhibiting cultural relic related to Zhaojun in Hohhot, such as inscriptions of "Tomb of Concubine Wang Zhaojun", "Green Mound of Zhaojun", "Mound of Concubine Wang Zhaojun", "Leave a Reputation Beyond the Great Wall" and "Coward Ashamed" and poems praising Zhaojun.

Except this Zhaojun Tomb, it is said that there are dozens of tombs of Zhaojun at the foot of Daqing Mountain. Jian Bozan, China's famous historian said that "Why there are so many tombs of Zhaojun. It is obviously that the appearance of Zhaojun tombs reflects Inner Mongolian people's good feeling about Wang Zhaojun. They hope that Wang Zhaojun was buried in their hometown."

In order to promote the national spirit and build a harmonious society, Zhaojun Culture Festival has been held every year since 1999. Various interesting programs in the fair have attracted a large number of tourists from all over the world. Wang Zhaojun is an ambassador of peace, and a historic witness of unity of people of all ethnic groups. The legend of Wang Zhaojun will be remembered and passed down.

Mausoleum of Genghis Khan

Driving from the famous steel city Baotou in the north of the Great Wall, acrosss the rolling waves of the Yellow River, drive about 120km on the Ordos Plateau, you will arrive at Atengxilian Town of Ejin Horo

Statue of Genghis Khan

Banner. Driving about 15km southwest ward, you will come to a one place called "gander Aobaoshan", where the tomb of proud son of Heaven, Genghis Khan is seated.

The illustrious name and great achievements of the Genghis Khan are widely passed down from generation to generation. He is the pride of the Mongolian nation and the most respected idol. The reason was simple, in the 5,000-year history of the Chinese nation, the Yuan Dynasty established by the descendants of Genghis Khan have made great contributions to unity of multi-ethnic nationalities. Moreover, in 1276 AD before the demise of the Southern Song Dynasty, the great Mongolian empire had been established in Mongolian Plateau in 1206. The man who created this great outstanding was Genghis Khan.

Genghis Khan (1162-1227) is founder of the Yuan dynasty, also named as Temujin, tribe named as Bo-Er-Zhi-Jin-Shi ruled by the Qi-Yan, a famous Mongolian leader in history, and a great military strategist and statesman. The meaning of "Genghis Khan" is emperor or Khan.

Genghis Khan was born in Mongolian Plateau. His father Yesugei once was a tribal leader. Since the Qin and Han dynasties, Mongolian Plateau in northern China has been a place where the nations Xiongnu, Xianbei, Rouran, Turkic, Uyghur, Jiejisi and other tribes had ruled the roost in succession. In the era of Genghis Khan, many tribes with different languages, different types of economy and culture and different social economic development levels appeared on Inner Mongolian Plateau. At that time, the society was changing from the primitive tribal to the class society. New rising aristocracies were on the endless war and killed each other in order to plunder more slaves, livestock and horses, beauties and fertile pastures to possess and maintain greater rights and interests. The social productive forces had been severely damaged.

The scene described in the *Secret History of the Yuan Dynasty* was like this:

Stars rotate, tribes rise in revolt.

Who can sleep soundly in bed! All come to rob treasure and slaves.

The earth is rolling up, there are troubles everywhere.

Who can sleep soundly in bed! People are killing each other.

Before the rising of Genghis Khan, the Mongolian lived in a nomadic life with no fixed abode. Sparsely-populated tribes were scattered in the vast prairie, and tribes had no chiefs or kings.

A hero is nothing but a product of his time. Temujin, with the strong indomitable character and intelligence, unified Mongolian tribes and ended a long period of chaotic situation on the Mongolian Plateau through the suppression of wars and massacre among the tribes and clans by leading a group of like-minded and courageous warriers. In 1206, at the meeting of tribal leaders held in the source of Wonan River (now the Enen River of the People's Republic of Mongolia), conforming to the historical trend, Temu-

Mausoleum of Genghis Khan

jin was elected as "Genghis Khan" and established the Mongolia Empire.

There are many records in the Chinese and foreign history books about the mausoleum of Genghis Khan, but only book *Tribe History made by Imperial Order* compiled during the Emperor Qianlong of Qing Dynasty and *History of Mongolian Nomads* written by Zhang Mu clearly recorded the tomb of Genghis Khan seated in Ordos. After the outbreak of the Anti-Japanese War, to protect the coffin of Genghis Khan, in June 1939, with the approval of the Kuomintang Government, mausoleum members assigned representatives jointly with the Mongolian representatives to transfer the coffin of Genghis Khan from Ejin Horo (meaning "master's mausoleum" in Chinese) via Yulin - Yan'an – Xi'an to Dongshan Buddha hall in Xinglong Yuzhong County, Gansu Province, and then had been worshiped for 10 years. In the summer of 1949, the Chinese People's Liberation Army marched into the eastern Gansu Province, when the Kuomintang general Ma Bufang retreated, who ruled the Gansu Province at that time, the coffin of Genghis Khan was moved to the Ta'er Lamasery, Huangzhong County, Qinghai Province.

At the beginning of the foundation of New China, under the unanimous request of the Mongolian people, approved by the Central People's Government, in March 1954, the delegation of Inner Mongolia Autonomous Region headed for Ta'er Lamasery. On April 23, they took the coffin of Genghis khan to Ejin Horo. In 1955, the people's government allocated special funds for the building of new mausoleum of Genghis Khan, which covers an area of 40,000m^2 with the building area of 500 m^2. Since then, the Chinese government has repeatedly allocated special funds to expand and renovate the mausoleum. Today, the mausoleum of Genghis Khan is both unique in national architectural style and magnificent in scale. Since the completion of the new tomb of Genghis Khan, the palmers and the visitors from all quarters paid homage, reverence and reminiscence to the great Genghis Khan every year.

On March 12, 1982, approved by the State Council, the Mausoleum of Genghis Khan was identified as the second group of the national relic protection unit.

Five Pagoda Temple

Located in the southeastern part of the old city of Huhhot, the Five Pagoda Temple was built in 1727 and completed in 1732, which was named as "Cideng Temple" in the Qing dynasty. The lastabbot of this temple died in 1886. So far, the tall and beautiful Five Pagoda Temple has been well-preserved.

The Five Pagoda Temple was built with masonry structure, with the convex shaped modeling. It is about 16.5m in height. The lower part is the throne of King Kong, and the top is five square pagodas. The throne of King Kong has 7 layers with the height of 7.82m. Of the five square pagodas, the central tower is slightly higher than others with the height of 6.62m. The others are lower with five floors. A large number of exquisite effigies and the expression of different gilt Buddha statues are preserved in these towers. Among them, there is the Buddhist scriptures "Jingang Boreboluomiduo Jing" ("Diamond Sutra" for short), which was written in Mongolian, Tibetan and Sanskrit. It is particularly worth mentioning that the north side of Bodhimanda relic pagoda is a brick screen wall, on which there are three pieces of line engraving stones. The central one is "Distribution area of Sumeru Mountian Xuni Hill", the west is a "map of the six great divisions in the wheel of Karma", and the east is an "astronomical map" with the diameter of 1.5m, marked by Mongolian with the zodiac the ten Heavenly Stems, 24 solar terms, 360-degree positions and names of Twenty-eight lunar Mansions. According to Mongolian inscriptions, the site map was carved according to the astronomical map made by the Imperial Board of Astronomy in 1725 AD. This is so far the only one astronomical map in China that used the ethnic languages, which is a valuable historical relic.

The green city-Huhhot

Huhhot is the capital of the Inner Mongolia Autonomous Region, a beautiful city, which is located in central Inner Mongolia, south of the Yinshan Mountains, at the foot of Qingshan Mountian, with an average altitude of 1,000m. It is adjacent to Yellow River to the west, the lofty Daqing Mountain to the north, standing on the Tomochuan Plain with Heihe River in the south. Hu Lv-jin, who was Chile nation of Northern Qi Dynasty (AD 488 —?) created the well-known *The Ode to Chile*, which said "Over the earth hangs the sky like a huge yurt. Between the vast

Five Pagoda Temple of Huhhot

sky and the boundless earth, Flocks and herds appear as grass bends to wind." With a few words, he described vividly the true portrayal of grassland scenery in Huhhot.

Huhhot is located in the middle reaches of the Yellow River Basin, is one of the cradles for the growth of Chinese nation. For thousands of years, China's northern Huns, Xianbei, Chile, Turkic, Khitan, Nvzhen and Mongolia nomads had camped and dominated here. In the long-term communications and integration with Central China in the long historical process, many profound historical footprints were left. For instance, the Great Wall and "Cloud City" (now in Togtoh County of Huhhot), which were constructed by the King Wuling of Zhao during the Warring States period; Chele plain, which was built by Xianbei nationality of the Northern Wei Dynasty; "Baidao chuan", which was built in the Sui and Tang dynasties; "Fengzhoucheng" (this site is in the eastern outskirts of Huhhot Baita, one of eight sights of Huhhot), which was built by Qidan people after the establishment of the Liao Dynasty. As times passed by, the prosperity of acient "Fengzhoucheng" city has been obliterated, leaving only a 43m-high "Wan bu hua yan jing ta" (Baita). The Baita has experienced years of wind and frost, but still stands proudly.

Modern Huhhot City are composed by two old

Night Scene of Huhhot

cities which were built in the different dynasties. The "Old City" was built in Ming Dynasty in 1581, its real name was the "Kuku he tun".The "Old City" was built by the Mongolian leader Alatan Khan and his wife and was named by the Ming Government as "Guihua". The "New City" was built in 1739 in the east of the "Old City" and got the "Shuiyuan" by the Qing Government. " In 1913, the two cities were incorporated as Guisui County, and became Suiyuan Province in 1928, and 'Guisui' City was built here. In 1954, the Central People's Government revoked the organizational system of Suiyuan Province and incorporated it into Inner Mongolia Autonomous Region. "Guisui"City changed its name to "Huhhot", which means a green city, symbolizing unity, youth and prosperity.

The surrounding cultural and historical relics are well-preserved, such as tomb of Zhaojun, White Tower of Liao Dynasty, the kiln cultural sites, Muslim temples, the Five Pagoda Temple, Xiaozhao Temple, Dazhao Temple, Princess House, House of the Generals, Yu-quan Well, Xilitu Zhao, Wusutu Zhao Monastery, Lama Cave, etc. After refurbishment, they all show charms in former days.

On July 31, 2007, the new eight sights of Huhhot were elected, including Zhaojun Museum, Five Pagoda Temple, Dazhao Temple, White Tower, Museum and former residence of comrade Ulanhu, Ely New Industrial Park, the Islamic Architectural Landscape Street and Ha Su Sea. It has not only ancient history, but also modern civilization. With social development and progress, constant innovation is required. Today's story will eventually become tomorrow's history.

Chapter 7
Landscape Regions of China's Grassland

I come from the grassland
Written by Ji Zhengmin, composed by Wang Xingming

When playing the Morin Khuur, larks are singing, and relatives gather together.
The melody with deep feeling of a long separation.
I come from the grassland.
I come from the grassland.
The grassland likes a profusion of flowers.
When Vega is rising, a pony is sucking the breast.
I will use my lifelong love to pay back the deep love of my mother.
I come from the grassland.
I come from the grassland.
There is so much to look forward to.
Whether I am like winds drifting to somewhere,
Whether I am like winds drifting to somewhere.
The grassland is always with me.

Grassland is one of constructural units of terrestrial ecosystem. In a special space dominated by grassland vegetation, that is to say, in a natural grassland region, nature is not completely homogeneous, with other ecosystems to compose a whole green vegetational cover by influencing each other, and these systems play an important role in ecological functions.

Taking Xilin Gol valley in Inner Mongolia as an example, the grass steppe is the sight background of this area, occupying more than 83% of total area of land, meadow and swamp account for about 6% or so, others such as sands forest and bushwood for 7%, and forest only accounts for 0.3%. So in a word, Xilin River is a typical steppe ecotope where perennial xerophytic grass steppe is dominant, the harmony between grassland and other soil types can supply natural resource base for the development of animal husbandry.

From a global perspective, China's grasslands belong to holarctic vegetation region, Asian central grassland subregion (main body) in Eurasian grassland region, Black Sea -Kazakstan grassland subregion (small part) and alpine steppe subregion of Qinghai-Tibetan Plateau (whole). The Asian central grassland subregion consists of Inner Mongolia plateau grassland, the Northeast China Plain grassland and Loess plateau grassland. The alpine grassland subregion of Qinghai-Tibetan Plateau consists of only one subregion. Xinjiang Altai Mountains upland grassland belongs to the Black Sea-Kazakhstan subregion.

Bunchgrass Grassland Ecoregion in Inner Mongolia Plateau

The bunchgrass region in Inner Mongolia plateau takes up the central part of Asian central grassland subregion. The geographical range consists of Inner Mongolia Plateau, which is joined with Mongolian grassland and outer Baikal grassland in Russia in the north and is bounded with the Yinshan Mountains in the south, including the easternmost forest steppe in this plateau, central typical steppes and western desert steppes. The altitude in the northeast of Hulun Buir is 600~ 800 m, and 1,000~1,200 m in middle part, and 1,300~1,500 m in the south. The eastern part is connected to forest ecotope of Daxinganling. The climate has such character-

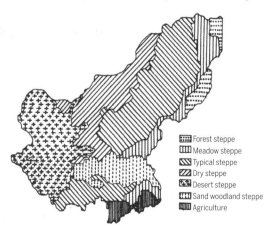

Ecological Division Map of Xilin Gol League, Inner Mongolia (quoted from Li Bo Memoir)

Chapter 7 Landscape Regions of China's Grassland 129

Landscape of Formation Stipa of eastern upland plain in northern Mongolia (Photographed by Yong Shipeng)

Eagle is flying in the sky (Photographed by Yong Shipeng)

The New Virgin Land (Photographed by Yong Shipeng)

Landscape of Northern foot of Yinshan Mountain, Hill and Highland (Photographed by Yong Shipeng)

Landscape of Xilin Gol Typical Steppe (Photographed by Xu Zhu)

Agricultural Landscape of Northern foot of Yinshan Mountain florescence of Flax and sunflower (Photographed by Yong Shipeng)

Landscape of Farming-pastoral Region in Xilin Gol Grassland (Photographed by Yong Shipeng)

Landscape of Farming-pastoral Region in Xilin Gol Grassland (Photographed by Yong Shipeng)

istics as follows: semiarid, cool in summer, and windy all the year. The annual average temperature in the east is -2~2 ℃, annual accumulated temperature above 10 ℃ is 2,200~2,500 ℃, and annual precipitation from east to west is 200~400mm. The soil is mostly chestnut soil and light chestnut soil. There is sediment sand bed on plain with different thickness, and smaller Tengri sand strip from east to west, bottom land and swamp around banks and lakes have some salinity soil. The annual precipitation in eastern part is 350~400mm with rhizome grass and weeds as natural vegetation, and *Pinus sylvestris* grows on sand of the northeastern region, and then trees and brush on the south dene. The central region is dominated by typical *Stipa* grassland represented by *Stipa grandis* and *Stipa krylovii* with an annual rainfall of 250~350mm, and dwarf under shrub and desert steppe distribute in the west with an annual rainfall of about 150mm. There are halophytic meadow and halophytic vegetation at the low-lying place or riverside in Inner Mongolia. The south of plateau had been reclaimed to farming-pastoral region, where the grassland is replaced by farmland.

Forest Steppe Ecoregion in the Northeast Plain

This region is an ecoregion of plain that is closest to the ocean in the grassland subregion of Asian middle part, where hydrothermal condition is good, soil is fertile, and it is in possession of obvious semihumid grassland characteristics.

This ecoregion consists of Songnen Plain and

Landscape of Arxan Mountain (Photographed by Yong Shipeng)

Liaohe Plain in the northeast with an altitude of 150~200m, the two plains are embraced on three sides by mountains and the landform is formed in this way that north, west and east are higher than south and middle part. In the north, the annual average temperature is 0~4℃, annual accumulative temperature above 10℃ is 2,300~2,600℃, and frostless period is 120 days. In the south, the annual average temperature is 4~6℃, and annual accumulative temperature above 10 is 2,800~3,000℃. Annual precipitation is 400~500mm,

Tessellated meadow and wicker pitcher at South End of Greater Khingan Grassland (Photographed by Yong Shipeng)

Autumn Harvest (Photographed by Yong Shipeng)

and precipitation of some regions near to mountain is 550mm. It rains concentrated from June to the August which adds up to 65% of annual rainfall. Acidulous black soil formed by clay proluvium in Quaternary period mainly distributes in upland plain with an altitude of 280~450m. On the contrary, alkaline chernozem distributes in lower plain, and then saline-alkali soil and part calcareous soil in low-lying places. Elm woodland grows in sandy soil and meadows grow on black land, rhizomatous grass and subordinate grass on chernozem, and salined meadow on saline-alkali soil. Besides, the plain has some smaller lakes embraced by reed swamp and reed meadow. At present, existing problems in agriculture are as follows: soil fertility is decreasing, grassland resources are degenerated, water resources are polluted and the structure of agriculture is simple.

The wet land protection area networks have been built for small rivers and lakes in this area. Fish and wild bird resource are abundant. There are lots of commercial fishes such as carp, crucian, silver carp, bighead carp and catfish, etc. There are only 16 kinds of crane in the world, and six species of crane including red-crowned crane, grey crane, white-head crane, white-naped crane, whooper swan and cygnet in this area. The area of reed field is very large where paper, weave and building materials can be made from, and fishes lay eggs. Cranes take fishes as their foodstuff. It is worth noting that maintain good relationships among reed, cranes and fishes to protect ecological balance of nature. Since there are lots of medicinal materials on plain grasslands, how to collect and protect medicinal materials and how to turn wild species into domestic species become a direction for a diversified economy.

Warm-temperate Grassland Ecoregion in Loess Plateau

This area is the earliest developed grassland ecological area in Asian central grassland region (cultivation area) because of good hydrothermal condition. Geographic area of this region includes Yanshan Mountain in the north, the south of Yinshan Mountains, Tai-Hang Mountains, Guancen Mountains and Lvliang Mountain in the east and Liupanshan Mountain in the west. There are valleys and basins among the mountains. Except these mountains, most

Plastic film mulching terrace of Loess Plateau in West Gansu Province (Photographed by Yong Shipeng)

Loess Hilly and Gully Region (severe water and soil loss) (Photographed by Yong Shipeng)

Autumn harvest of river valley in southern Gansu Province (Photographed by Yong Shipeng)

part of this area belongs to Loess Plateau. The surface of plateau is divided into pieces and formed interlaced landform of tableland, bridges, loess hills and gulleys. There are island-like low mountains on plateau such as Huanglong Mountains, Laoshan Mountain and Ziwuling, etc. All these mountains are made of stone. The altitude of plateau is 1,000 m in general, or 1,500-1,800 m. The temperature varies with altitude and sites in the north and south. The annual average temperature, annual accumulative temperature over 10 ℃, and frostless period in the north are 6.5~9.0 ℃, 2 900~3 200 ℃, and 150~180 days, respectively. Comparatively, the annual average temperature, annual accumulative temperature over 10 ℃ in central south are about 6.5~9.0 ℃, and 2 900~3 200 ℃, and the corresponding data in the far south are 12~13 ℃, and 4 000~4 500 ℃. The frostless period in the south is 200 days. As for annual precipitation, it is about 450~500 millimeters in the east, and 200~250 millimeters in the west. Some warm-temperate zone crops which can't be planted in temperate zone such as apples, pears, Chinese dates, walnuts, grapes, peppers, cotton and winter wheat can be grown in areas with above-mentioned temperature or irrigation conditions. The soil of this area formed by loess texture is divided into different Dark loessial soils and grey brown soil from south to north. This area is called grassland region nominally, but actually is farming-pastoral region. Grassland area is not large at present and the plateau field has been opened-up and the residual grassland is very small.

This area may be divided into two grassland areas, regions-southeast forest steppe and northwest bunchgrass grassland area.

1. Southeast forest steppe area

With annual precipitation of 400~550 mm, the forest steppe zone is scattered with Aneuvolepidium chinense, Themeda triandra and seedy steppe on Dark loessial soils. The plateau surface and loess hills are reclaimed as farmland to plant drought-tolerance crops such as millet, oat, sorghum, potato, legume, wheat,

2. Northwest bunch grass grassland region

The annual precipitation is 200~250 millimeters. There is grayish brown on loess parent material. The native vegetation is *Stipa capillata*, formation *Stipa breviflora* and this region belongs to typical steppe zone of warm-temperate zone. There is desertification grassland in the westernmost. This region is a farming-pasturing region taking pasture as the principal thing. The severe loss of water and soil occured in the region because of seriously destroyed vegetation.

Since the animal husbandry-oriented policy, agriculture-animal husbandry and agriculture-oriented policy have been changed for more than 11 times over the past 20 years in this area, the meadow was destroyed tremendously. The productivity of residual meadow declined, amount of gramineous grass extremely reduces and rare sheep resources of Ningxia are degenerating gradually.

Alpine Steppe Ecoregion of Qinghai-Tibet Plateau

Qinghai-Tibet Plateau is located in the subtropics with an altitude of 4,000~5,000 m. The terrain rises from southeast to northwest. The moisture decreases from southeast to northwest because of the barrier action of the southern Himalayas and Nyenchen Tanglha Mountain, which prevents the Pacific Ocean monsoon especially moist air current of Indian Ocean from going into by making a detour at the Hengduan Mountains. The vegetation distribution on the plateau from southeast to northwest is alpine meadow brushwood, alpine steppe and alpine desertification in proper order.

This region is located in the southeast, middle part and northeast of Qinghai-Tibet Plateau, and divided into semihumid alpine forest meadow and semiarid alpine meadow of middle southern.

Yaks and Tibetan Sheep on Autumn Pasture (Photographed by Xu Zhu)

Alpine Kobresia Meadow Region

Taking watershed of Nyenchen Tanglha mountain as the southern boundary, this region consists of northwest of Sichuan, west of Gannan and east of Northern Tibetan Plateau, located in the zone of transition from alpine and gorge region at southeast of Qinghai-Tibet Plateau to the surface of plateau, and the northwest borders on central grassland region of Qinghai-Tibet Plateau. The climate characteristics are alpine-cold and semihumid, dry and less rainfall in winter and cool and moist in summer. The amount of annual rainfall is 500~700 mm and mainly rains from June to September. Thunderstorm and hails are frequent in summer.

There are high mountains and canyons in the southeast and the vegetations in shady canyon and sunny canyon are different. *Kobresia pygmaea* meadow at widely distributed in the northwestern hills and broad valleys, forming zona columnaris of *Kobresia* Willd. meadow. The vegetation form is comparatively simple, which is dominated by *Kobresia humilis* meadow.

Alpine *Kobresia* meadow distributes on the dale and slopes where the altitude is 4,000~5,200 m with good system of drainage, and mid-xeric plants which are low and dense are in the ascendant. The

Equus Kiangs (Photographed by J. Marc Foggin)

Kobresia pygmaea with the height of 3~5 cm is often seen, besides, there are *Kobresia yangii, Kobresia vidua, Kobresia royleana, Kobresia tibetica Maxim, Kobresia humilis, Kobresia littledalei* and *Kobresia utrculata*, etc.

Such meadow is characterized by short grass, simple species, monotonous physiognomy, short of shrubs and ephemeral spermatophyte. Lots of typical alpine plants with the traits such as dwarfism

Tibetan Sheep

(10~15cm), small leaves, few ramus from stems dispensed in this region, and some of them are rosette plants including *Polygonum macrophyllum, Pedicularis oederi, Pedicularis przewalskii, Meconopsis Racemosa, Anemone demissa, Lamiophlomis rotata, Trollius pumilis, Saussurea stella Maxim, and Leontopodiumnanum.* Cusion plants include *Androsace tapete, Arenaria musciformis* and *Arenaria kansuensis* and succulent plants consists of *Sedum dubium*, etc.

This meadow grassland are with good traits that the roots are dense, turf is solid and elastic. So this meadow is a good natural pasture and it is suitable to graze cold-resistance livestock such as yaks and Tibetan-sheep.

The agricultural region distributes in mountainous region and valley terrace covering a small area, staple crops include highland barley, oat, pea, Yuangen, potato, radish and oil seed rape, etc. The crop distribution is about an altitude of 3,500m in the north and 4,000m in the south. Yak and Tibetan sheep are the main animals. There are rare animals such as *Crossoptilon auritum, Crossoptilon crossoptilon, Phasianus colchicus,* and *Cervus elaphus*, etc. in the forest bush of canyon area, and some rare medicinal materials such as *fritillary, Rheum palmatum, Lagotisclarkei, Saussurea involucrata* etc as well. Some animals such as *Bos mutus, Equus hemionus,* Tibetan antelope, *Cervus albirostris, Tetraogallus tibetanus* are on the Alpine meadow and there are some others famous medical materials such as *Gentiana macrophylla, Corispermum hyssopifolium* and *Figwortflower Picrorhiza Rhizome*.

Alpine Grass Steppe Region

This area consists of western part of Himalayas in Qinghai-Tibet Plateau, most part of Qiangtang Plateau and southern part of Qinghai Plateau. It is cold, high and semi-arid in this region with alpine steppe soil. The parent materials are accumulating efflorescence crust of weathering and debris such as granite, sandrock and conglomerate, etc. The vegetation on river valley and plateau is represented by alpine steppe. The southern part has arcticalpine meadow, constant green bushwood and pulvinoid vegetation due to semihumid climate. *Gramineous Stipa* is the main grassland vegetation, except for some special species belonging to Qinghai-Tibet Plateau. A lot of species belong to temperate and warm-temperature zone.

Taking Gangdise Mountains-Nyenchen Tanglha Mountain as boundary, this region can be divided into two different grassland sub-regions. The northern sub-region is composed of alpine grassland and alpine desertification grassland, and the south is composed of alpine grassland and alpine meadow. The temperate zone grasslands and dry bushwood appear in Yarlung-Zangbo River Valley.

1. Alpine Steppe Southern Sub-region (alpine bush-grass grassland and shrub grassland sub-region)

It is located between southern Himalayas and Gangdise Mountains-Nyenchen Tanglha Mountain.

There is big difference in air temperature between the valley and high mountain of this sub-region. The average temperature of the valley is 5~8 ℃ with an altitude of 3,500~4,000 m, while the average temperature of mountain on both sides of valley is from 0~ -3 ℃ or lower with an altitude of 4,600~5,500 meters. With altitude rising, the surface of area is covered by snow and ice. In this way, the vegetation types in this region are different with different landform and altitude.

Vegetation in the south of Tibet Plateau is mainly composed of *Stipa purpurea* and *Stipa breviflora*. Alpine meadow is dominated by Kobresia humilis and creeping shrubs grown on the shady and sunny slopes, respectively.

The upper reaches of the Yarlung Zangbo River including the broad valley between Himalayas and Gangdise Mountains have formed alpine steppe mainly consisting of *Stipa purpurea*. *Caragana versicolor* is scattered on the north Slope of Himalayas, *Kobresia humilis* distributes in some places where the altitude is 5,200~5,400 m and the *Stipa Glareosa*, *Artemisia wellbyi* desert steppe are extensively distributed on the alluvial plains where the altitude is 4,600~4,700 m.

There are lots of resources of medicinal plants such as the *Saussurea tridactyla*, *Rheum palmatum*, *Nardostachys jatamansi*, *Lamiophlomis rotata*, *Gentiana algida*, *Corydalis boweri Hemsl*, *Lagotisclarkei*, etc. There are rare animals such as *Equus hemionus*, Tibetan Antelope, *Bos mutus* and so on.

Xeric *Sophora moorctoftiana* in Tibet and *Ceratostigma griffithii* minor are extensively distributed on the valley and hills of middle reaches Yarlung Zangbo River with an altitude of 4,000 m. *Stipa bungeana*, *Stipa capillacea*, *Orinus thoroldii*, *Pennisetum flaccidum* grassland distribute on places with an altitude of 4,000~4,500 m.

The valley region boasts abundant sunshine, fertile soil, high air temperature, good irrigation conditions, and abounds in wheat and highland barley, as "granary of Tibet".

2. Alpine steppe Northern Sub-region (alpine bush-grass steppe sub-region)

It is located in the north of Nyenchen Tanglha-Gangdise, including Yushu of Qinghai Province, Naqu of Tibet and east of Ngari Prefecture.

The alpine grasslands growing *Stipa purpurea* and *Stipa pinnate* and *Carex tristachya* are widely distributed in hills, hillside, proluvial fan and plateau in with good drainage and an altitude of 4,500~5,300 m.

Mountainous Bunch Grass, Dwarf Forb Desert Steppe Ecoregion in Xinjiang Altai

China's vegetation is divided into eastern and western grassland sub-regions.

The western grassland sub-region is located in the northernmost of Xinjiang, consisting of west of Altai Mountains, Sawuer Mountain, Tarbagatai Mountains and inclined plain areas as well as Wuerkashar Mountain. The range of this subregion from north to south is very narrow. It is the easternmost from Black Sea to Kazakhstan grassland subregion in the southwest of Eurasian grassland to China, and it is a grassland subregion transiting from Altai abundant area to Mongolian grassland community.

Altai Mountains possess smooth terrain with multilevel peneplain. Because it is located at the center of Eurasia, the climate is dry due to the influence of dry climate from Zhungar desert and Zhaisang desert. The total rainfall over the year is about 200~300mm. The altitude of mountainous regions is about 2,000~3,000 m and the relative height is 1,500~2,500 m. The soil and vegetation have obvious vertical zoning phenomenon.

China's western grassland sub-region is transiting to eastern grassland sub-region. In the zonal transitional community types, both vegetation types from Kazakhstan grassland and Mongolia grassland can be seen. For example, some species such as *Stipa*, *Stipa sareptana*, *Pennatae*, *S.kirghisorum*, *S.lessingiana* and *Festuca sulcata* can be often found not only in Kazakhstan grassland but also in Mongolia grassland.

In the mountainous regions, the typical vegetation vertical structure takes upland meadow as baseband, and the band spectrum is as follows from bottom to top: upland meadow zone, mountainous coniferous forest or mountainous retama bushland zone, subalpine meadow zone, and alpine meadow zone. In the northwest of humid Altai Mountain, there are some changes in the belt, where subalpine meadow belt is replaced by the subalpine deciduous broad-leaved shrubs and alpine meadow is also replaced by the alpine tundra.

The vegetation here is complex in genetic classification system because it is located in regions where there are components of Mongolia and Europe-Kazakhstan, ephemeral plants and desert vegetation from Gobi Junggar, and thus the transitional nature is very obvious.

Altai western grassland is divided into two vegetation regions: Altay grassland region and Daerbaertai - Sawuer Mountain grassland according to different hydrothermal conditions.

Stipa of coniferous forests, *Festuca sulcata*, ephemeral forbrich upland meadow region in Altai Mountain

This region extends from the northern foot of Sawuer Mountain in the south, east to upriver of Eerqisi in Altai Mountains, and to national boundaries in northeast and northwest. The climate of grassland is cold and wet. The annual rainfall of the foot of piedmont plain is 165~190mm, and over 600mm in mountainous area, so water system of this region is relatively dense. Eerqisi River, the third largest river in Xinjiang is on the plain.

The mountainous vegetation of this section is desert steppe with *Stipa glareosa* as constructive species from piedmont plain where the altitude is 800 meters to lower mountain belt where the altitude is

1,500 m. The bushveld belt above can be found in the sunny slope where the altitude is 2,100 m and larch forest appears on the shady slope where the altitude is 1,200 m. Mingled forest of larch and Siberian spruce appear on the belt with an altitude of more than 1,500 m. Subalpine meadow has not been developed extensively because of shading by coniferous forest of in the northwest corner of Altai Mountains, but subalpine meadow is widely distributed in the northern mountain of Altai Mountains.

The most prominent characteristic of grassland vegetation combination is that *Stipa glareosa* communities dominate in the center of Asia located in desert steppe are superior in numbers, but communities that composed by *Festuca sulcata*, *Kazakhstan Stipa* and *Artemisia gracilescens* also distribute in this region. In addition, the large formation of *Stipa subsessiliflora var.basiplumosa* is still the dominant species in bushveld.

Altai grassland is the most important animal husbandry base in Xinjiang. The efficient use area of natural grassland accounts for 1/5 of total area in Xinjiang, the stocking rate is 1/4 of Xinjiang. At present, the stocking rate is only about 10%, so it has great potential to develop animal husbandry in Altai area.

The main shortcoming of this region is that the area of grassland in winter is so small and only accounts for 18% of the whole grassland area, so the pasture area in four seasons is unbalanced. The pasture area in summer is surplus while short in winter.

Festuca sulcata, *Artemisia*, Ephemeral forbrich steppe in Tarbagatai Mountains-Sawuer Mountain

Several mountains in the western Junggar Basin consist of Tarbagatai Mountains, Sawuer Mountain and Wuerkaer Mountain, which have 4~5 peneplain ladders with the highest altitude of about 2,000 m, some peaks could reach 2,500~3,000 m. The peneplain at different levels maintained intact and the terrain slopes gently.

Although the mountains are affected by the desert climate, there are still much moisture as the Arctic ocean wet airflow can reach here. The annual precipitation is more than 280mm and piedmont plains develop into typical desert steppe.

The seasonal distribution of annual rainfall is relatively even. Taking Tacheng meteorological data as example, the precipitation in spring is 69.6mm, 84.2mm in summer, 81.5mm in autumn and 45.0mm in winter. The rainfall of spring and early summer accounts for more than one third of total rainfall, thus it provides conditions for the upgrowth of ephemeral plants in spring.

The average annual temperature is 6℃ in this area; the average temperature is 22℃ in the hottest month, the coldest month average temperature is -12~-14℃, ≥10℃ accumulative temperature is 2,800℃, which are climate characteristics of typical temperate grasslands.

The vertical structure vegetation of from top to bottom is: desert steppe zone typical steppe-shrub steppe with *Spiraea Salicifolia, Parochetus communis* shrub zone - subalpine meadow with *Rosa acicularis* shrub with thorns- alpine *Kobresia* and *Carex tristachya* meadow zone.

The desert steppe takes *Festuca rupicola* and *Stipa* as constructive species, and *Artemisia lessingiana, Artemisia sublessingiana, Artemisia gracilescens* as coedificator. It is completely different from Eurasian grassland region which is dominated by *Stipa glareosa* and *Stipa gobica* communities. Another feature of vegetation in this area is that there are lots of ephemeral plant species in grassland community such as *Poa bulbosa var.vivipara, Ferula dissecta, Tulipa schrenkii, Plantago minuta, Lepidium apetalum, Alyssum desertorum, Tetracme quadricornis, Ixiolirion tataricum,* the *Veronica, Tauscheria, Alyssum, Ceratocarpus* and *Schismus*.

In mountainous region with hard stones, the area of *Festuca rupicola* + *Stipa* is not large. They and shrub species form shrub-steppe communities. *Spiraea hypericifolia, Caragana frutex, Calophaca soongorica, Amygdalus ledebouriana, Atraphaxis frutescens, Festuca rupicola, Stipa capillata, Stipa kirghisorum* and *Stipa macroglossa* together form stable shrub-steppe communities, which can be hardly seen in other areas of grasslands in China.

In mountainous region, shrubs are widely distributed. The scrub composed of *Rosa acicularis, Lonicera tatarica* and *Spiraea hypericifolia* distribute from low mountain belts to subalpine bottom. The bushwood is accompanied with lots of mesophyte plants such as

Piedmont Steppe

Polygonum glabrum, Phlomis tuberosa, Dactylis glomerata, Poa pratensis, Bromus inermis, Helictotrichon asiaticum and *Origanum vulgare*, etc.

Subalpine forbrich steppe in Tarbagatai -Sawuer grassland develops well and occupies the region with an altitude of 1,800~2,100m. The dominant species include *Erodium stephanianum, Aconitum soongoricum, Dracocephalum nutans, Ligularia altaica, Trollius altaicus*, Smooth Brome, *Bromus inermis, Alopecurus*

Populus euphratica forest

soongoricus and *Poa pratensis*, etc.

Alchemilla Spp. meadow is widely distributed in the lower part of subalpine. This meadow is relatively short, but has many species.

Alpine meadow is a typical example of Asian Central Mountain and has not been developed in a large area due to the short mountain. Communities composed of *Kobresia smirnovii* and a variety of *Carex tristachya* distributes on the highest peneplain and occupies a small area.

Tarbagatai-Sawuer grassland is one of the main animal husbandry bases of Xinjiang. The pasture output of four seasons is balanced. In order to further develop animal husbandry, it is necessary to improve pasture of natural grassland in cold season and establish artificial grassland.

Tacheng, Emin and Hebukesaier counties are one of the major grain-producing areas in Xinjiang, with a higher agriculture management level. In recent years, a large area of land has been reclaimed, among which peneplain ladders of low hills have been changed into farmland, yielding relatively stable hawests and the harvest is relatively stable.

Grassland regions are lack of forest resources. A small area of Siberian larch and spruce forest are scattered in Sawuer Mountain, while other mountains only have scattered poplar woods.

Chapter 8
Sustainable Utilization and Management of Grassland Resources

Mongolian Home on the Grassland
Written by Si Uyunqimg, composed by Le Balashazhaer, Translated by Buren Bayaer

Wild deer and cattle are sleeping
Onto the vast grassland is my mother's sight
The sweet fragrance of hot tea is redolent with irresistible smell
The Mongolian grassland is looking into the distance
Deer and livestock share the meadow
Mongolian homes on grassland are enjoying the peace
In the sunlight camels and sheep and awakening
The sun rises from first glimmer of dawn
The milk fragrance spread from the rooftop-window of Mongolian yurts
Deer and livestock share the meadow
Mongolian homes on grassland are enjoying the peace

The grassland is a renewable natural resource with ecological, economic and social functions. It is an important production base of animal husbandry and the gene treasury of biological diversity. It has some special functions such as keeping ecological balance, adjusting climate, preserving water and soil, wind prevention and sand fixation. However, overgraving, blind reclamation and predatory management have resulted in the "black dusters" on grasslands. Therefore, for a sustainable development of humankind, it is necessary to carry out scientific research on natural grasslands and integrated control on degraded grasslands.

The grassland is a biogeographical type formed in the process of natural and historical evolution between forest and desert. The grassland ecoregions form major continents with a certain ecological convergence, and are the evidence to the parallel development of some ecosphere on earth. The grassland biosphere has established material basis for human evolution and development. Friedrich Engel said that "the appearance and development of the grassland opened a new chapter in human history." When humankind learned to use fire, they also learned how to produce and use production tools, so their activities hindered grassland ecosystem

Historical Experience and Lessons of Exploitation and Utilization of Grasslands

Overgrazing (Photographed by J. Marc Foggin)

Water and soil loss in mountainous region (Photographed by Yong Shipeng)

Grassland desertification (Photographed by Yong Shipeng)

Grassland stony desertification (Photographed by Yong Shipeng)

Water and soil loss (Photographed by Yong Shipeng)

and surroundings more and more. The pristine nomadism had a little effect on the ecological environment, thus the grassland remained ecological balance and became one of cradles for human development.

With the development of the society, the increase of population and domestic animals, especially large-scale cropland reclamation and overgrazing of grassland regions have brought great pressure on ecological environment of grassland regions and fragile ecosystem. So the original vegetative covers dominated by grass and leguminous plants had been replaced by some secondary plants such as artemisia and forbs. Large-scale land reclamation turned grasslands into farmland, which was discarded due to the poor soil, making the ecosystem of grassland severely damaged and the normal energy flow and logistics order no longer existed. At all times and all over the world, the memory of global environmental disaster resulted by improper exploitation is stoll fresh. In former days, Babylon was one of birthplaces of ancient civilization, but now fertile grasslands on the lands of Babylon have become the barren due to endless reclamation and overgrazing, feeding 1/4 people compared to Hammurabi period. From 1930s to 1950s, black storm occurred in the plains of American and western Canada, and West Siberia-northern Hazakhstan Plain which shocked the world. For example, on May 12, 1934, extremely severe sandstorm occurred on the prairies of the American and the western Canada, spreading 24km in length from east to west and 400km in width from north to south, which nearly covered two thirds area of America. This extremely severe sandstorm takes away about 300 million tons of surface soil from west coast to east coast, and the yield of winter wheat decreased by 5.10 million tons compared to 10 years ago.

In recent years, sandstorm occurred frequently in northern China, affecting provinces and regions along south and north of Yangtze River. Except natural factors such as global climate warming and reduced precipitation, the other factors such as overgraving and excessive reclamation also destroyed grassland vegetation, resulting in the grassland degradation. For the sake of historical reasons or reality factors, ecocatastrophe caused by grassland degradation has set off alarm bells.

Inscription of Premier Zhu Rongji (Photographed by Shan Guilian)

Road Etching (Photographed by Yong Shipeng)

In 2002, China's former premier Zhu Rongji clearly indicated that it was extremely urgent to contol desertification and build green barriers when he visited the southern Xilin Gol League. Moreover, he also pointed out the direction to further build the grassland ecological environment. We must protect the grassland ecological environment so that the human environment could be protected. On contrary, if we destroyed the grassland ecological environment, the nature must punish us. So we should remember the historical experience and lessons.

Current Situation of Grassland Degradation and Comprehensive Prevention Countermeasures

At present, the worsening of desertification around the globe has threatened the survival of humankind and caught the world's attention. The grassland degradation is the major reason of the land desertification in the arid and semi-arid areas worldwide.

There are two reasons for grassland degradation: one is global warming, climate warming, and the reduction of precipitation and grass yield. The other is over grazing, which leads to the overload of grasslands. This result is that the normal energy flow and circulation of materials in ecological system of grasslands have been destroyed, the ecological functions exhausted and the construction broken down, so the environment is deteriorated, the productivity of land descended and the primeval vegetation is replaced by secondary bare land.

For this purpose, we introduce ten countermeasures about integrate control of degraded grassland for your reference.

1. Implementing compensated use of grasslands and establishing accounting system of the grassland resources

For a long time, because the right use of grasslands has not been established, the grassland resources have been utilized by herdsmen at random. It is one of main factors for grassland degradation. Inner Mongolia Autonomous Region government takes the lead in implementing of household contract responsibility system and compensated use of grassland, and it has received good

Contrast test of grassland improvement (Photographed by Yan Zhijian)

Grassland Improvement (Photographed by Yan Zhijian)

Artificial Grassland (Photographed by Bartel)

effect. So we suggest that the compensated use and right to use of grasslands should be established throughout the country, and the value calculation and pricing system for national grass resources should be developed. The grassland resource accounting should be included into Grassland Law. We should Develop, use and manage grassland resources according to law as soon as.

2. Actualizing grassland improvement, positively constructing the grassland and increasing grass and livestock

At present, China's pasturing areas and farming-pastoral regions are at the end of depending purely on natural grassland raising livestock because the number of livestock has exceeded the bearing capacity of nature grassland. We must adjust grazing capacity and improve the quality of grassland to relieve the pressure of grassland. Mu Us Sandland, Erdos of Inner Mongolia builds family farms to make local herdsmen get rid of poverty and become better off. China has various grassland types, so the measures to increase grass should be taken according to circumstances. In the northern grasslands, it is important to center on meadow steppes with better water and soil conditions. For example, Daxing'anling-Xiao Xing'an Mountains-Lvliang Mountains-Liupan Mountain present zonal distribution, and borders between farming and ranging regions, the area of which is 400 thousand km^2. This zone can be built into the largest production base of red meat and furs in China's northern pasturing area and farming-pastoral region.

3. Reforming the species of livestock and raising the production performance

In some regions with good conditions, we can improve the production performance and product quality through reforming the species of livestock. As a result, we can control the quantity, raise economic returns, and lessen the grassland pressure. Finally, we can achieve the purpose of preventing grassland degradation.

4. Increasing input

Governments at all levels must increase investment in the range build-up, control the grassland degradation, improve the productivity and increase income of the farmers. According to statistics, the governments spent about RMB100 million into the grassland construction from 1991 to 1995, and the average input is 0.45 yuan per hectare of grassland. The herdsmen cannot help but facing the low input and low output. So it is suggested to set up the grassland construction fund, reinforce the investment in range build-up, and encourage herdsmen to invest in construction and improve the productivity.

5. Strengthening the legal management and earnestly implementing Grassland Law

Some ruthless exploitation, overgrazing and coyoting shall be put to an end. We should focus on unordered reclamation for meadow steppes. Overgrazing is not permissible on arid and semi-arid typical steppes, desert steppes and alpine steppes, and the number of livestock on the grass must be strictly controlled. Moreover, Through improving livestock and feeding models, we can improve the output of livestock products. We should protect the nature and the wildlife in the extreme arid Gobi Desert. Some grassland may be built into national parks or natural protection areas to satisfy the needs of biodiversity protection, ecological tourism, education and scientific research.

6. Controlling the population of the pasture areas

In general, the pasture area holds a vast territory with a sparse population. According to statistics in 1991, the density of population in Inner Mongolia is 18.7 per square kilometer, in Xinjiang is 9 per square kilometer and in Qinghai is 6.1 per square kilometer, much less than interland. But according to calculation of land bearing capacity, it is overcrowded especially in the severely destroyed areas, where people's life standard is very low. According to the third population census data, the growth rate of population in the pasture areas is between 2.5 ‰ and 3.4 ‰, which are much higher than the national average speed of growth.

7. Improving the scientific quality of the grassland administrators

It is necessary to train the grassland administrators and basic decision makers, create the idea of the resource value and sustainable utilization of grasslands and put prevention of degraded grassland into practice. So we can establish training and educational systems at different levels.

8. Reinforcing efforts in returning farmland to grassland and improving the level of ecological construction

In the farming-pastoral areas and pasturing areas, extensive cultivation exists in crop farming, especially in the desert and subtemperate regions. It is wise to return farmland to grassland on inferior plots where input and output are imbalanced. From the perspective of the government, it must increase the grassland area, gradually expand cover area, and enhance ecological protection efficiency.

Because of vast grassland territorg and diverse conditions, in different areas, ecological construction cannot use one model, and we must adjust measures

Table 8-1 Main Natural Grassland Resources in China

Grassland type	Grassland subtype	Distributing region	Optimum domestic animals
Temperate grassland	Meadow steppe	Northeast plain, the eastern Inner Mongolia plateau	Dairy-beef dual-purpose cattle, fine-wool sheep,
	Dry steppe	Middle Inner Mongolia plateau	Meat cattle, fine-wool sheep, half fine-wool sheep
	Desert steppe	The middle western of Inner Mongolia plateau	Fine-wool sheep, half fine-wool sheep
Warm-temperate grassland	Upland thicket	Northern China low hill	
	Upland herbosa	Northern China low hill	
	Typical steppe	Northern China loess plateau	Fine-wool sheep
	Desert steppe	Northwest loess plateau	Fur sheep
Alpine steppe	Meadow steppe	Southeast Qinghai plateau	Poephogus grunniens, tibetan sheep
	Typical steppe	Middle of Qinghai plateau	Poephogus grunniens, tibetan sheep
	Desert steppe	Central northern Qinghai plateau	Poephogus grunniens, tibetan sheep
Temperate desert	Steppified desert	Northwest Inner Mongolia, north Xinjiang	Camel
	Typical desert	Western Inner Mongolia, north Xinjiang	Camel
	Extreme desert	Southern, middle, western Xinjiang	Wild Camel
Alpine desert	Steppe desert	Central northern Qinghai plateau	Tibetan-sheep
	Typical desert	The north of Qinghai plateau	Wild hoofed animal
	Extreme desert	Qinghai plateau north and alpine zone	
Meadow	Alpine lowland meadow	Southwest alpine	Poephogus grunniens
	Upland grassland slope	Southern China hills	Buffalo, cattle
	Lowland meadow	Everywhere	Cattle
	swamp	Everywhere	Not suitable for use
Open forest grassland	Open forest grassland	Forest zone	Dairy-beef dual-purpose cattle, reindeer

according to local conditions.

9. Expanding the effective utilization of crop straws

China has abundant crop straw resources. In Inner Mongolia, the annual output of cornstalk and straws in half farming-pastoral Leagues and counties is 300 thousand tons. But a small portion was used to feed the livestock because they were burned as fuel. Increasing the utilization of crop straws can not only increase rural income and promote the adjusting of industrial structure, but also change the situation about livestock husbandry development relying solely on grassland. It can lessen the grassland pressure and avoid the grassland degradation.

10. Strengthening scientific research on grassland and developing ecological monitoring on grassland

It needs to deepen the research and study on

grassland degradation mechanism, control biological and nonbiological factors' influence in grassland degradation, and offer some theoretical foundation for the recovery and management of degraded grassland. The grassland is a dynamic ecosystem and keeps changing like the terrestrial ecosystem. So we must know the change tendency in time and develop the ecological monitoring on grassland.

Ecological Principles and Technical Strategy of Reasonable Utilization of Grassland

The natural grassland renewable resource with many ecological functions and economic values as thenatural forest. On one hand, it offers fodders for the herbivores, on the other hand, it is the home to various rare wildlife, pests, rats and their natural enemies, and also the gene bank for biological resources such as fine pasture, valuable Chinese herbal medicine and edible mushrooms, etc. The grassland has specific ecological mechanism, which can form soil fertility, conserve water and soil, improve the microclimate, and maintain the ecological balance of surrounding environment. We must protect the grassland and utilize grassland reasonably so as to create material wealth and exquisite environmental conditions.

Rational utilize and preserve grassland resources

Develop modern animal husbandry base with regional advantages according to the natural distribution law of the grassland resources.

Like natural forest, the natural range resource is affected by atmospheric circulation and monsoon climate as well as the tectonic structure and geomorphy, and has some remarkable zonal difference and regional characteristics. The grassland resources under different natural historical and geography background own special grassland ecological system, biota, food chain network, growth rhythm, nutritional ingredient and the environment characteristics, which are basic conditions to form local fine livestock species. Many fine livestock types such as Sanhe horses, Yili horses, Yanqi horses, Hequ horses, Sanhe cattle, Yili cattle, Tianzhu yaks, Alashan red camels, white cashmere goats, Altay Fuhai sheep, Xilin Gol sheep, Ningxia tan sheep, Xinjiang fine-wool sheep, have been developed in special grasslands under long-term natural selection and artificial selection. It can be said that different natural grasslands are cradles of different qualified livestock species. China's temperate grasslands are fit for feeding the Mongolian sheep, horses and cattle, the semi-desert temperate grassland is suited to feed tan sheep in Ningxia and set up top grade fur commodity

Artificial Grassland Chipping Pasture Photographed by Bartel

base; the Qinghai-Tibet Plateau alpine steppe and alpine semi-desert steppe are suitable for feeding yaks and Tibetan sheep; and arid and extremely arid desert steppes are suitable for establishing camel industry. To introduce the excellent foreign varieties, we must consider the biological rule about the livestock adapting to the local environment.

The modern animal husbandry can get a sound development on the basis of the correct zoning. But the correct ecological regionalization is on the basis of comprehensive analysis of the distribution rule of the grassland resources. We must carry out comprehensive survey and evaluation on grassland resources on a regular basis, soas to work out plans about the utilization of national grassland resources and provide scientific basis for the modern animal husbandry.

Determining the quantity of livestock according to grass growth

The quantity of livestock should be determined according to grass growth based on level of productive forces of natural grassland green plants.

The animal husbandry development is governed by grassland ecological rule. Forage grasses provided by different grassland types have some limits and changes.

Having known the grazing capacity of a grassland region, we can adjust the development scale and speed of stock breeding according to the dynamic change of the herbage production. If the number of livestock falls short of normal threshold level of the grassland, it indicates that the grassland resources have some potential. Otherwise, if the number of livestock is close to or exceeds threshold level, it needs to adjust the development scale or seek to improve the grassland primary productivity and execute the management system of determining the quantity of livestock according to grass growth.

Storing up enough basic forage grass and adjusting the balance between grass and stocking

China's primary productivity level of every natural grassland resource has great disparity between the seasons and years except the imbalance of zones. From the ecological environment of the grassland livestock, periodic bioclimatic catastrophic events occur everywhere including drought, snow disaster, wind disaster, disaster caused by hail, fire disaster, insect disaster, rat disaster, and epidemic, etc. In addition to adopting specific countermeasures, we should adapt to natural fluctuation rules of primary productivity of grassland resources to achieve the modernization of animal husbandry. We should ensure the sound and stable development of animal husbandry and adjust the balance between grassland and livestock according to the strategy of "making up for a crop failure with a bumper harvest".

In China's temperate area, the seasonal imbalance of the primary productivity of natural grasslands exist objectively, and annual fluctuation is a normal natural phenomenon, which is another objective limiting factor affecting the stable development of animal husbandry.

Developing orderly management operating system on the basis of evolutionary changes of natural grassland

The natural grassland is a natural dynamic balance system. When it is not disturbed by external force, it will develop and change according to natural law, and achieve the stage of climax. A large area of range resources of China's temperate grasslands, warm-temperate desert steppes, alpine desert steppes and alpine steppes of Qinghai-Tibet Plateau are biogeographical environmental resources at natural succession climax community stage. When people graze, mow and reclaim on grasslands, they will exert pressure on grassland ecosystem. When the pressure is within an adjustable elastic range, the system will keep the normal condition about the construction and function. But on the contrary, when the pressure exceeds the adjustable limit, the grassland will show retrogressive succession and the grassland environment will transform into non-grassland environment, which will result in serious consequences such as soil and water loss, land desertification, soil salination and swamping. So we must develop scientific grassland utilization system according to natural succession law of natural grasslands.

We will discuss the relation between the grazing succession, mowing succession, reclaiming succession of grassland and scientific grazing system, mowing system and system of returning the grain plots to husbandry in the abandoned land.

1. Grazing succession and establishing the grazing system

Grazing is a normal food chain relationship between the domesticated animals and plants on the grasslands and it is the essential condition to keep the normal construction and function of grassland. Once grazing is absolutely prohibited, a lot of fine pasture will be replaced by cheesy pasture, which provides breeding condition for many noxious animals. However, frequent ingestion and endless trample will definitely destroy the ecological functions of grassland and lead to the loss of population diversity, structure simplification, the land solid or excitation (sandy soil). So the grassland renew is difficult and then the grassland environment is getting worse, which give rise to retrogressive succession of grassland.

The grassland retrogressive succession is a gradual change process of grassland environmental degradation. When the grazing pressure exceeds the bearing capacity of living beings, the grassland system will have qualitative worsening, and then the normal succession will be replaced by retrogressive succession. It can be seen from sere table of temperate grassland grazing that the different life forms of plant population have regular alternate and replacement with the increase of grazing capacity. Therefore, we can assess the grassland successional stage according to the indicative plants on the pasture and set out forecast in time, so we can take the active strategy and prevent the reversal succession. For the degraded grassland, we can take following measures: adjusting the stocking construction and lessening the grazing pressure; banning leisure grazing; rotation grazing; opening up new source of water and moving the place to graze; and ploughing the soil and transforming permanently.

2. Mowing land succession and establishing mowing system

Clipping the dry grass can help to solve the unbalance between stocking and grass in non-growing seasons. At the same time, the unremitting mowing at regular intervals is the existing condition for some special grassland types (such as beach meadow and rhizome grass steppe). Once the mowing activities are stopped, these grasslands will be replaced by others. For instance, the beach meadow will be replaced by brush and the rhizome grass steppe will be replaced by bunch grass steppe. It is necessary to establish the reasonable mowing system which can ensure the existence and development of fine mowing land. However, the unreasonable mowing, such as mowing for many successive years and mowing at improper time will cause the degradation of vegetation and grassland, and retrogressive succession of grassland system (Table 8-2).

Steppe of *Aneurdepidium chinensis* is one of the best mowing lands in China's eastern area of temperature grassland. The cured hay made up by *Aneurdepidium chinensis* boasts a high reputation at home and abroad. At present, the steppe of *Aneurdepidium chinensis* with unreasonable mowing system has degraded, and some sections with alkaline spotting exposed barely provide good-quality hay.

Table 8-2 Sere of the Temperate Grassland Grazing

plant type community / grazing succession stage / The signal climax	Meadow steppe	Typical steppe	Desert steppe
Normal grazing stage	Subterranean grass + mesophytic forbs (*Aneurolepidium chinensis, Bromus inermis, Sanguisorba officinalis., Medicago falcate* and so on)	Tussock grass steppe (*Stipa grandis, Cleistengenes, Koeleria cristata, Agropyron cristatum, Bupleurum falcatum, Siler divaricatum* and so on)	Small bunchgrass steppe (*Cleistogenes songorica, Cleistogenes songorica,* and *Lagochilus ilicifolius,* etc.)
Light grazing stage	Subterranean grass + Subterranean Carex spp.+ xerophytic forbs (*Aneurolepidium chinensis, Carex duriuscula, Vicia sepium* and so on)	Tussock grass+ artemisia community (*Stipa grandis, S.krylovii, Artemisia frigida*)	Small bunchgrass+ artemisia family steppe (*Cleistogenes songorica, Artemisia frigida* or *Artemisia xerophytica, Ajania fruticulosa* and so on)
Heavy grazing stage	Subterranean grass + Subterranean Carex spp. + mesic xerophytic forbs (*Carex duriuscula, Aneurolepidium chinensis, Thermopsis lanceolata* and so on)	Artemisia + Micro grass community (*Artemisia frigida, Cleistengenes, Carex duriuscula, Heteropappus altaicus*)	Artemisia family steppe (*Artemisia frigida* or *Artemisia xerophytica, Ajania fruticulosa*)
Over grazing stage	grazing tolerance forbs +Carex lanceolata (*Iris lactea var.chinensis, Carex duriuscula, Stellera chamaejasme* and so on)	Inferior quality forbs +Artemisia community (*Heteropappus altaicus, Artemisia scoparia, Chenopodium album* and so on)	Inferior quality forbs community (*Peganum nigellastrum, Convolvulusammanii, Artemisia pectinata*)

Unreasonable mowing, that is to say, continuous mowing in the same district for many years, the height, coverage, weight and relative dominance of fine pasture in the grass group reduced in general, and the diversity of flora component decreased, the ply structure simplified and as time passes, reverse replacement of community occurred. To protect the natural mowing pasture, we must establish reasonable mowing system, formulate rotation mowing plan and make sure the plants on mowing pasture can rehabilitate; for soil hardening, it is a good method to improve the soil ventilation status and the utilization ratio of the surface runoff through shallow ploughing and loosening the soil; apply fertilizer in some conditioned places in rainy season, and supplement the fertilizer source to improve the yield and quality of the mowing.

In the extensive arid, semiarid and alpine pasturing areas, the precipitation is less, natural environment is severe, so we should grow fine pasture artificially according to local conditions, increase the coverage of green plants covering and supplement the insufficient natural grass.

3. Abandoned land succession and establishing the system of returning the grain plots to pastures

It is a long history of reclaiming meadows and growing crops. While developing single crop farming in grassland regions will gain a certain amount of food in rainy years, the seeding rate is greater than harvest yield in dry years. As time goes on, a lot of fine grassland was abandoned as it is not suitable for planting, which is common in China's northern pastoral-farming areas.

The abandoned land will gradually transform from pioneer plant communities to secondary grassland once human intervention is stopped. This succession is named vegetable recovery succession of the abandoned land or positive succession. The diversity and structural complexity of plant community will push ahead after experiencing a stage, and finally it will achieve the native grassland entering a new stable stage. Under normal circumstances, the temperate grassland and bunchgrass steppe will recover to the native grassland taking about 15~20 years. If the succession was interfered by the other activities such as grazing, it will always remain in the lateral branch deviating from normal sequence, what is called "disturbance community" (Table 8-3).

Table 8-3　Effect of continuous mowing for many years on the Northeast *Aneurolepidium chinensis* Grassland

Utilization Degree	Community Name	Total Coverage (%)	Fresh Forage Yield (0.5kg /667m^2)	Plant Species (m^2)	Proportion of high quality grass (%)	*Leymus chinensis*		Forbs	
						Coverage Proportion (%)	Weight Proportion (%)	Coverage Proportion (%)	Weight Proportion (%)
Contrast	*Aneurolepidium chinensis*, *Stipa* baicalensis, forbs	85	560	16	77.3	60	59.6	35	26.0
Continuous mowing and light degradation	*Aneurolepidium chinensis*, Tussock grass, forbs	80	530	22	60.0	30	24.0	50	51.3
Continuous mowing and heavy degradation	Tussock grass, *Aneurolepidium chinensis*, forbs	75	528	18	36.0	20	12.0	45	64.5

The vegetation recovery in the abandoned grassland will have four stages, which is summarized as follows.

> Closed cultivated land
> ↓1~2 years
> Annual, biennial forbs community
> ↓2~3 years
> Rhizome grass stage (*Aneurolepidium chinensis* or *Aneurolepidium secalinus*, *Elsholtzia ciliata*)
> ↓5~10 years
> Rhizome grass — bunch grass stage
> (*Aneurolepidium chinensis*, *Stipa*)
> ↓15~20 years
> Bunch grass stage (secondary *Stipa* steppe)

The graph above shows that the tame native grassland has to experience a long time and several stages to recover to the native grassland; in the course of grassland restoring succession, it needs to reach the rhizomatic grass rehabilitated stage before achieving the stable bunch grass steppe state. The superior species usually are *Aneurolepidium chinensis* with high feeding value at this stage. People can use this rule to prevent or defer the succession of abandoned land, and then the rhizomatic grass stage can be stored. Following measures can be adopted to make land transformed to the semi-artificial grassland with higher economic value.

(1) Leave land uncultivated with seeds of *Aneurolepidium chinensis*, and promote the course of rhizomatic grass stage and increase the density of fine forage.

(2) Seed the *Aneurolepidium chinensis* meanwhile

increasing sowing the seeds of leguminous forage (*Astragalus*, *melilotoides ruthenica*) which can improve the nutritive value of forage and the soil fertility.

(3) After entering the later period of rhizomatic grass stage, we can loosen soil to create good soil conditions for rhizomatic grass. Through these measures, we can establish a set of supervising system to the lands returning cropland to grass land instead of negative land abandonment.

Grading Criterion of Grassland Degradation

The reason of grassland degradation is that grassland ecosystem is disturbed by exterior factors such as overgrazing, and then the normal orders of energy flow and cycle of matter are destroyed, which have resulted in environmental degradation, reduction of productivity of land, gradual decreasing of ecological functions, structure disorganization, and native grassland changing into secondary bare land, etc.

The grading standard of grassland degradation is developed based on the diversity of grassland types and differences in driving force of grassland degeneration in various regions. We will take the specific composition of various grasslands, ecological groups, laminar structure, regeneration characteristics, level of primary productivity and the structure of food chains as objective indicators to classify the degradation degree of grasslands.

The grades of grassland degradation are divided into four levels: light degraded, moderate degraded,

Light degraded grassland (Photographed by Shan Guilian)

Medium degraded grassland (Photographed by Shan Guilian)

Heavy degraded grassland (Photographed by Shan Guilian)

heavy degraded and extremely degraded, and they are marked as green, yellow, orange and red early warning signals (Table 8-4).

Table 8-4 Grading Criterion of Grassland Degradation

Degradation Degree	Floristics	Epigean biomass and cover degree (%)	Ground cover and surface situation	Soil situation	System structure	Restorable degree
I. Light degraded	There are no important changes in primary community. The quantity of dominant species decreased, palatable species disappeared or decreased	Drop by 20~35	Ground cover reduced significantly	No obvious change, and hardness increased slightly	No obvious change	Resume quickly after enclosing
II. Moderate degraded	Constructive species and dominant species substitute each other, but most of initial species remained	Drop by 35~60	Ground cover disappeared	Soil hardness doubled. There is erodent mark on the ground. Salt quantity increased in low and swamp soil	Carnivores decreased and herbivores rodent animals increased	Resume quickly after enclosing
III. Heavy degraded	More than half initial species disappeared. Species composition is simple. Short and tramping-tolerance grasses are dominant	Drop by 60~85	Bare ground	Soil harness increased twice as much as before, the content of organic matter decreased remarkably, and alkali spots appeared	Food chain shortened significantly, and the system structure was simple	Resume difficultly and improvement measures needed
IV. Extremely degraded	Vegetation disappeared or only sparsely grass	Drop by more than 85	Bareness or alkali spots	Loss value in use	System disaggregated	rebuild needed

It is the simplest way to use indicative plants to identify the degree of grassland degradation. At the primal state of temperate grassland in our country, constructive species of grassland community are composed of *Stipa baicalensis*, *Stipa grandis*, *Stipa klemenzii* and *Stipa bungeana*, etc, together with some companion species, such as, *Bromus inermis*, *Melilotoides ruthenica*, *Vicia*, etc. The meadow steppes have *Astragalus melilotoides*, *Bupleurum*, *Heracleum*, *Glycyrrhiza*, etc. When grazing intensity is increased, the number of these grass seeds at primal community gradually dropped off, and then disappeared at the heavy-degraded stage. The number of other components such as *Cleistogenes squarrosa*, *Atemisia Frigida*, *Thymus serpyllum*, *Serratula chinensis*, *S.Moore* and *Potentilla acaulis* increased with the grazing intensity, even turned into dominant species and then formed degraded communities. At desert steppes, *S.kelemenzii* (*Cleistogenes songorica*, *Stipa glareosa* and *Stipa breviflora*) were constructive species, and then formed short grass steppe. If grazing intensity was increased, the degree of dominance of *Stipa klemenzii* declined, and *Cleistogenes songorica* became dominant species, and then harmful weeds such as *Artemisiapectinata*, *Convolvulus ammannii*, *Peganum harmala* appeared in grass community, it showed that grassland entered into the heavy-degraded stage.

Thus it can be seen from that the indicative plants to distinguish the degradation degree vary in different natural terrains and grassland types. Only when you find out the succession laws on grassland, the indicative plants can be used to evaluate the degradation degree on grasslands. The grassland degradation caused by such actions as digging up and mining can be evaluated according to stabilization of soil base and sandy flow. The degradation degree caused by rats and insects can be evaluated according to the dense of mouseholes or insects. As for soil and water loss degree on Loess Plateau, the degradation degree should be confirmed by corrosive modulus. Establishing the target system on grassland degeneration is a quite complicated technology. In order to carry out monitoring work on grassland, many technical projects have been carried out in China, which is used as major scientific basis.

Management of Grazing Land on the Grassland

Grazing refers to feeding behavior of domestic animals on grasslands. Grazing is the main measure of grassland management and the main approach for domestic animals production. Domestic animals have much effect on grasslands by picking up grass, trampling and excretion:

(1) Domestic animals pick up forage grass selectively. When food consumption of some grass is no more than 50% of plant material, it will have little effect on grass growing, and facilitate the tiller, ramification and grow of grass. But if grazing frequency or grazing period is irrational, pasture that tastes good will decrease or disappear from grassland, on the contrary, some grass that tastes bad and some poisonous grass will increase.

(2) When domestic animals walk or run on grasslands, they will destroy herbage and surface soil. The long-term and excessive trample on grassland will make grassland ground become bare, and permeability of soil declined, and then cause water and soil loss. Rational grazing and proper trample can crash mulch which formed by ground moss and algae, and it is good for herbage seeds to grow, and it can raise organic content of soil by accelerating dead plants to break up and decompose.

(3) In the course of grazing, faeces and urine of domestic animals are the nutrition of pasture. Adult cattle with the weight of 500kg can excrete nitrogen of 7.5kg, phosphor of 3kg and potassium of 4kg. Grazing can make herbage and domestic animals supply nutrition for each other, and facilitate circulation of materials in grassland ecosystem. But the density of livestock and excessive manure will pollute herbage and have a bad effect on grassland.

Irrational grazing will make grassland plant

community degraded, the production performance of grassland decline both qualitatively and quantitatively, and then restrict the development of grassland livestock husbandry and destroy grassland environment. In order to prevent grassland from degeneration or destroyed caused by overgrazing, it is necessary to take scientific measures to manage natural pasture.

1. The quantity of livestock should be determined according to grass growth and strictly control grazing capacity of grassland

The number of total domestic animals have tripled from 29.16 million to 92.00 million since the establishment of the People's Republic of China (PRC). But the carcass weight of a yak declined from 400kg in 1950s to 100kg now in Qinghai province, and the carcass weight of a sheep declined from 40kg to 20kg. It is thus clear that though the number of domestic animals tripled, the quantity and quality of livestock go downtrend. Therefore, we should change the practice of immoderately pursuing the increase of livestock and grazing capacity blindly on grasslands. The quantity of livestock should be determined according to grass growth.

2. Implement plot of rotational grazing by using advanced technology

Plot of rotational grazing refers to that grazing land is divided into some plots according to their productivity and the characteristics of use. Implementing designed rotational grazing can not only feed livestock well, but also keep the productivity of pasture.

Compared to continuous grazing, rotational graz-

Rotational grazing (Photographed by Yan Zhijian)

ing has following advantages:

(1) Using grassland economically and effectively and then increasing grazing capacity and the quantity of animal by-products. Rotational grazing can support more than 30% livestock with the same area of grassland, so it can increase more than 35% productivity of livestock.

(2) It can change composition of grassland vegetation, and improve yield and quality of herbage. It is reported that fine pasture increased from 64~69 to 303~305 plants per square meter, and the grass height increased from 28~36 cm to 78~86 cm.

(3) It is convenient for the management and improvement of grassland, and it is also in favor for carrying out responsibility system of grassland management and production responsibility system of animal husbandry.

Rotational grazing is a highly technological work. We should firstly consider cycle and frequency of rotational grazing and plot size, which are the basis to work out rotational grazing plan.

Rotational grazing (Photographed by Xu Zhu)

3. Adjusting measures to local conditions, allocating species of livestock

The types of grasslands have obvious regional characteristics, which made livestock have formed relevant living habits and skills to adapt themselves to the ecosystem. For example, cows and horses like eating tender and succulent tall grasses; sheep like eating some short grass and artemisia species that are soft with more dry substance; camels and goats pick up some shrubs and undershrubs which are thick, rigid, hard, thorny and olfactory with higher ash content; yaks and Tibetan sheep like eating dense and low grass. In order to exert fully potential productivity of all kinds of ecosystems, it is necessary to divide grassland into plots according to regional characteristics of ecosystem and adaptability of livestock, and adjust

Horses on Sunit Grassland (Photographed by Yong Shipeng)

Cattles on Grassland at southern foot of Yinshan Mountain (Photographed by Yong Shipeng)

Cattles on Grassland (Photographed by Yong Shipeng)

Sheep are eating on Achnatherum splendens steppe (Photographed by Yong Shipeng)

measures to local conditions to arrange species of livestock. In China's northern grasslands, cows are in east, sheep in middle, camels in west, and yaks and Tibetan sheep are in southwest, which are the best grassland ecosystems as a result of long-term efforts according to different natural environment. Once the grassland ecosystems are destroyed or change a lot, the species of livestock will change accordingly. If Sanhe cattle are raised at Qinghai-Tibet Plateau, then it is different for them to adapt; also, it is different for yaks to live on Inner Mongolia Grassland. Therefore, only optimally allocating species of livestock optimally, can the potential productivity of grassland and livestock be brought into full play, and grassland ecosystem keep optimum balance.

4. Making use of growth vigor of pasture to develop seasonal animal husbandry

The best period of growth for pasture is from June to September, during this period, rainwater is abundant, herbage is dark green and enrich in nutrition and the domestic animals are plump and sturdy. Making full use of herbage growth of the best season, we can feed more livestock and realize the aim that livestock are strong in summer and fat in autumn. Before winter comes, we can avoid such situation by weeding out and slaughtering animals and alleviating the pressure on pasture in spring. This is a scientific and economic way of managing fatstock, which has been proved effective in some places in China.

5. Confirming reasonable grazing period according to developmental state of herbage

The period from the beginning of suitable grazing to the end of grazing is called the period of grazing or the season of grazing. During this period, grazing pasture is destroyed the least and benefited the most. Grazing too early or too late is harmful to grassland and livestock. If grazing begins too late, palatability and nutritive value will be decreased and regeneration capacity will be weakened, and this will affect on the time of reuse. If grazing ends too early, the grassland will not be used fully and resources will be wasted; and if grazing ends too late, there is not enough time for perennial grass to store nutrient, and then have bad effect on the yield of the next year.

Chapter 8 Sustainable Utilization and Management of Grassland Resources 163

White cashmere goats of Erdos (Photographed by Yong Shipeng)

Goats on sandy grassland (Photographed by Yong Shipeng)

Fine-wool sheep of Jirem (Photographed by Yong Shipeng)

Domestic Cervus elaphus (Photographed by Yong Shipeng)

Management of Mowing Pasture of Grassland

Mowing pasture, also known as clipping pasture, is only for mowing grass but not for grazing. Haymaking is one of the important keys of operating management of grassland, and it is important for storing up forage grass and decreasing the death of livestock caused by short of forage grass. At present, there are about 2,000,000,000 m^2 of cutting grassland in China, though it is only 5% of the total area of grassland. They can supply complementary feeding in winter or basal feed for pasturing areas and rural areas every year.

Mowing has an important effect not only on grassland plants and soil. But mowing affects the botanical contents of grassland community and yield of aftergrass. The quantity of the herbage whose regeneration capacity is weak will decrease because of cutting; on the contrary, the quantity of the herbage whose regeneration capacity is strong will increase. Mowing also has an effect on the structure of soil and fertility. Mowing will affect the growth of pasture as dry branches and fallen leaves decrease, and soil become harden after rainfall.

Establishing a reasonable mowing system is a necessary condition to ensure the sustainable utilization of grassland. Continuous and improper mowing will make grassland degenerated. In order to prevent the grassland from degeneration, the period of rotational cutting should be confirmed by means of combing holidays with breeding according to local weather conditions and features of herbage growth. In this way, the growth and breeding of herbage can be accelerated and then the cutting grassland will be used sustainably.

1. Establishing cutting time according to local phenological period of herbage

If mowing is too early, the quality will be good but the yield will not be high and, if cutting too late, though more dry matter can be gained, the quality will decline. Based on the growth characteristics of forages, the appropriate cutting time of gramineous forage grass is at heading date. If Legumes are the main grass, the appropriate cutting time is at flowering period. It is better not to cut during rainy days in order to prevent new stubble from decomposing which will affect regeneration of herbage.

2. Taking rotational cutting system, alternate cutting and grazing

Rotational cutting means that grassland is divided into several plots and then clipping period and frequency are changed according to a certain order so that herbage can store much more nutrient substance and yield seeds. In this way, it will be convenient for grassland plants to renew and propagate, improving plants growth conditions, and then realizing alternant cutting and grazing.

3. Establishing cutting height according to regeneration traits of grass

Height of stubble has an effect on the yield and regeneration. If stubble is too high, the branch will be produced from node stem of herbage and such branch does not has its own independent root system, so this situation will affect yield and quality of herbage. If stubble is too low, it will hurt growing point of grass and then make regeneration capacity reduced. As for natural grassland with more bottom grass, the appropriate height of stubble is about 3~4cm. For rangeland with more top grass, the appropriate height of stubble is about 5~6cm.

4. Fertilizing scientifically to facilitate high and stable yields of herbage

Apply fertilizer on cutting grassland can significantly increase yields of herbage, and control composition of grassland vegetation. If possible, it is suggested to use organic fertilizer, which is complete fertilizer containing not only three elements such as nitrogen, phosphor and kalium, but also a lot of organic matter. After organic fertilizer is decomposed, it will contribute to form soil granular structure and then increase the capacity of water holding and fertilizer maintenance. The aftereffect of organic fertilizer can last for a long time.

5. Generalizing mechanical mowing, increasing labour efficiency and quality of hay

As mowing costs a lot of time and labor strength is strong, mechanical mowing is very important in forage grass clipping in China's pasturing areas. Hay mowers used at present can facilitate the mowing and follow-up work. For instance, hay combine-harvester can make clipping, fattening, ridge and sun-cure completed in one step. It has been found that mechanical cutting not only increases labour efficiency, resolves the problem

Chipping pasture on rangeland (Photographed by Bartel)

Peak season of Leymus chinensis mowing pasture (Photographed by Xu Zhu)

Mechanical mowing (Photographed by Xu Zhu)

Hay bundle of Leymus chinensis (Photographed by Yong Shipeng)

of labour shortage in pasturing areas and harvests good grass in time, but also decreases the cost of mowing.

Management of Resources of Medicinal Plants on Grassland

Herdsmen are mowing grass (Photographed by Bartel)

Grassland is one of the main production bases for medicinal plants in China. There are a lot of species of medicinal plants on grasslands, which are extensively distributed with a large number of reserve and distinctive geographical characteristics. The exploitation and utilization of medicinal plants are a part of rational utilization of grassland resources and realizing industrialization of grassland. With increasing market demand for medicinal plants, wild medicinal plant resources have been severely damaged and the ecological environment of producing areas has been deteriorated due to predatory digging. Owing to the demand surge for herbal medicine, about 20% natural medicinal plants are endangered. So we give suggestions as below.

1. Establishing a benign ecological pharmacy development mode

In order to avoid damaging ecological balance of grassland in the process of herbal medicine production, we should guide the production of Chinese herbal medicine with the concept of ecological balance and following economic laws. Establishing a benign ecological pharmacy development mode means that combining agriculture, forest and stockbreeding with medicine respectively, in this way, Chinese traditional medicine and Chinese herbal medicine resources can achieve synergetic development.

2. Establishing a medical plant garden for rare and endangered species

Owing to higher economical values and the increasingly expanding contradiction between need and supply, lots of rare medical plants resources such as *wild Panax ginseng, Gastrodia elata, Panax Pseudoginseng Wall. var. notoginseng, and Cordyceps sinensis*, etc. have been damaged and are endangered. Introduction and domestication of rare and endangered medical plants on grassland, ex situ conservation and turning wild into cultivation are effective measures to protect medical plant species. Social and economical benefits can be gained and endangered degree can be reduced to a certain degree by studying biological and ecological features and generalizing scientific research results.

3. Realizing the protection for Traditional Chinese Medicine resources and modernization of sustainable utilization

With the help of modern biotechnology, introduction, domestication and artificial cultivation of the medicinal plants on rangeland can enlarge species and quantity of natural resources of Chinese medicinal materials and improve the quality of Chinese medicinal materials. The pre-warning system for rare and endangered medical species and the storage of resources should be developed. Chinese medical resources should be managed and protected scientifically and used rationally with the help of modern production technology and resource management.

4. Effectively protecting medical plants resources of rangeland by taking comprehensive countermeasures

In order to protect the resources of medicinal plant on rangeland as a part of whole natural resources, it is necessary to take comprehensive countermeasures, which include matching laws and regulations and administrative systems, coordinating technical measures and economical measures, encouraging preservers of wild resources, punishing destroyers of wild resources, and restricting users of wild resources. Protection zones for rare and endangered species of medical plants shall be built up. Preservation technology for endangered species shall be developed, and medicinal plant germplasm resources shall be collected.

We should study on assessment system of germplasm characteristics and the technology of ex situ conservation and in vitro conservation, and build up gene pool of rare and endangered species.

Besides resources of medicinal plants, the rangeland also have plentiful of flowering resources such as delphinium grandiflorum which has the unique shape and looks like the blue flying swallow on the branch when come into flower; and two colored gmelin sealavender herb which has densest flowers, white calyx and yellow corol. In addition, there are others beautiful and rare plants such as *Dianthus chinensis*, *Gypsophila* and *Haplophyllum dauricum*, etc. These beautiful flowers will be widely used under the impact of using native land flowers and plants to beautify landscaping.

Systemically Developing Grassland Dynamic Monitoring and Realizing the Sustainable Utilization of Grassland Resources

Setting up a protection zone of conservation of natural resources is an important symbol to test a country's modern scientific and technological level. Under high attention of Chinese Communist Party and government, we have established state natural resource management system, and have done a lot of work in the conservation of forest resources and rare and wildlife. The scientific investigation and orientation study also have some foundation. Although some natural protection zones have been built in temperate grasslands, there is no national protection zone of grassland resources which can embark on study on the structure and functions of grassland ecosystem, on protecting grassland diversity, on surveying dynamic changes of grassland and increasing grassland productivity and on keeping grassland ecological balance of different areas. So we chose several regions listed below as national protection zone of grassland resources:

① Natural Conservation Zone of the Salt Marsh Rangeland in Northeast Plain;

② Natural Conservation Zone of Hulun Buir Meadow Steppe of Inner Mongolia;

③ Natural Conservation Zone of Xilin Gol Typical Steppe of Inner Mongolia;

④ Natural Conservation Zone of Altai Upland Meadow in Xinjiang;

⑤ Natural Conservation Zone of Upland Meadow in Xinjiang Tianshan;

⑥ Natural Conservation Zone of Qilian Mountain Upland Meadow in Gansu;

⑦ Natural Conservation Zone of Alpine Steppe in Qinghai-Tibet Plateau.

Other than general protection zones, the natural protection zones of grassland refer to comprehensive management and protection of range resource in specific natural regions.

Now several questions about survey on grassland ecological monitoring are introduced.

"The conception of "grassland ecological monitoringe

Grassland ecological monitoring is a foundation to rationally use grassland, to scientifically manage grassland and to predict ecosystem of grassland. Investigation and observation on special steppe communities and whole biological communities of grassland ecoregions (including plants, animals and microorganisms) and habitat (including weather, soil, hydrology and the geochemical process) and their relationships are required, and the speed and dynamic process of energy flow and circulation of materials in grassland ecosystem should be explored to provide relevant information to grassland management department to make up relevant countermeasures.

The "grassland ecological monitoring" takes ecological monitoring network as main base positioned at ground level, and combines with satellite remote sensing technology of landsat.

The content of "grassland ecological monitoring"

1. Natural dynamic law of grassland ecosystem

The natural and dynamic law of grassland ecosystem includes seasonal changes, annual changes and long-rang variations, which include the following:

The natural and dynamic laws are consisted of contents as follows:

① The relationship between the productivity of grassland vegetation and climate fluctuation;

② The development observation of plant population in grassland communities;

③ Chemical composition of plants and its dynamic assessment;

④ Dynamic observation of rodent population in grassland biocommunities;

⑤ Dynamic observation of main insect population in grassland biocommunities;

⑥ Dynamic observation of soil animal population in grassland biocommunities;

⑦ Dynamic observation of edaphon population in grassland biocommunities;

⑧ Observation of steppe soil withered grass and litter;

⑨ Observation of soil physical properties and chemical property of grassland;

⑩ Dynamic monitoring of hydrological status of grassland.

2. The evolution trends of grassland ecosystem under human disturbance

The variation trends on grassland ecosystem under human disturbance are mainly comprised of following:

(1) Retrogressive succession.

① The trends of degenerated grassland caused by overgrazing or mowing;

② The course of soil desertification (including

Astragalus membranaceus (Photographed by Yong Shipeng)

Glycyrrhiza uralensis (Photographed by Yong Shipeng)

Scutellaria incana Biehler (Photographed by Shan Guilian)

Dianthus chinensis (Photographed by Shan Guilian)

Gypsophila paniculata (Photographed by Yong Shipeng)

Haplophyllum dauricum (Photographed by Shan Guilian)

Gentiana scabra Bunge (Photographed by Shan Guilian)

Coix lacroyma-jobi (Photographed by Yong Shipeng)

Paeonia lactiflora (Photographed by Yong Shipeng)

Herbal Ephedrae (Photographed by Shi Wengui)

Delphinium grandiflorum (Photographed by Yong Shipeng)

Limonium sinense (Photographed by Xu Zhu)

the influence of reclamation);

③ Grassland pollution caused by pesticide, fertilizer, industry and mining waste water and exhaust gas;

④ The effect on grassland ecosystem by exploiting mineral resources such as coal and oil, and building of thermal power stations;

⑤ The effect on grassland ecosystem of urbanization;

⑥ The effect on grassland ecosystem by firewood chopping and collecting and digging up domestic fungus.

(2) Positive succession.

① The effect on grassland ecosystem by improvement measures (such as enclosure, fertilization and irrigation);

② The effect on grassland ecosystem by growing grass, forestation and plant feed;

③ The effect on grassland ecosystem by improving livestock.

Several opinions on "grassland ecological monitoring"

① The "grassland ecological monitoring" takes different ecoregions as objects. The overall arrangement of ground monitoring points shall make the focal points stand out and give consideration to general information, and data and information obtained should be typical and representative.

② The grassland ecological monitoring should exactly illuminate dynamic variation rule.

③ The grassland ecological monitoring should have a long-term plan and be persistent, otherwise, it is difficult to achieve expected goal.

④ Emphasis should be put on observing the variation trend of grassland under the influence of human activities.

⑤ The normalization of grassland ecological monitoring, facility modernization, networking of distribution of survey platforms and data systematization should be realized step by step. When conditions are mature, the book named *Technical Manual of Ecological Monitoring on National Grasslands* should be compiled and issued by relevant department of the state.

Postscript

At the time of publishing this book, a poem of *Farewell on the Ancient Grassland* written by Bai Juyi, a poet in Tang Dynasty fell a reverie. It is just like the ancient oracle makes the poem for the modern grassland. According to the explanation of *Detailed Analysis of Three Hundred Tang Poems* by Yu Shouzhen, the original intentions of writers is to express poetic feelings by metaphor, which have a profound message, called eternal words. Therefore, they have been passed down for thousands of years, and widely read by people all the time.

Farewell on the Ancient Grassland
Bai Juyi
Lush, lush grass on the plain,
Once every year it sears and grows.
Wild fire can not burn it out,
Spring breeze blows it back to life.
Distant scent invades the ancient path,
Sunny green joins the arid towns.
Another send-off to out wanderlust son,
Sad, sad leave-taking exuberant.

Today, let us borrow this poem to express feelings for grassland. We hope that we can add some new meanings to this ancient poem, and as the tag of this book.

"Lush, lush grass on the plain, once every year it sears and grows." It shows a piece of picture of vast grassland in China, which is brimming over with vigour and vitality. This is a natural phenomenon from withered to lush, year in and year out and go round and round.

"Wildfire can not burn it out, spring breeze blows it back to life." It reveals the secrecy of grassland from ancient times to today. The grassland has indomitable vitality. Whenever Spring returns to The Good Earth, the grassland will revitalized, bringing a piece of verdancy to the ground.

"Distant scent invades the ancient path, sunny green joins the arid towns." When people begin to care about grassland with concrete actions, the weeds may grow quickly, just like thousands upon thousands of horses and soldiers, approaching to desert, and changing the desert into oasis.

"Another send-off to out wanderlust son, sad, sad leave-taking exuberant." Everything in this world will go through meetings and partings. In the beginning of the heaven and the earth genesis, mankind have forged permanent relations with grasslands, it may be called live and die together. However, a man's life is limited, but the life of grassland is infinite. Generation after generation follows close to another and shares the same blue sky with grassland.

We hope that humans live in harmony with grassland, and grassland is always flourishing.

Written in early winter in 2007, Huhhot, Inner Mongolia

References

A.A. Junatovii, translated by Li Jitong. 1959. Basic Characteristics of the Vegetation in People's Republic of Mongolia[M]. Science Press.

A.II, translated by Zhang Shen. 1959. The Meadow Vegetation of Russia[M]. Science Press.

A.N. Strahler, A.H. Strahler. 1986. Modern Natural Geography[M]. Science Press.

Ba Tu, Tai Lin. 1993. An Introduction to Economic Development of Animal Husbandry of China's Grassland [M]. Beijing: Nationalities Publishing House.

C.B. Cox, translated by Zhao Tieqiao et al. 1985. Biogeography[M]. Higher Education Press.

Chen Changdu. 1964. Where is the boundary of mid-section of Steppe sub-region (Erdos) in China[J]. *Plant Ecology* & *Geography Botany Periodical*, 2(1).

Chen Min. 1998. Research on Improving Degraded Grassland and Building Artificial Grassland [M]. Huhot: Inner Mongolia People's Publishing House.

Chen Wen. 1992. Economic Study of Grassland Stockbreeding. Huhhot[M]. Inner Mongolia University Press.

Dong Weihui, Hou Xixian. 2002. Grassland Murine Ecology and its Control[M]. Inner Mongolia People's Publishing House.

Du Qinglin. 2006. Strategy of Sustainable Development of China's Grassland[M]. China Agricultural Press.

E.M. LAFULIANKE *et al.* 1956. Russia Vegetation Overlay Directory[M] (the second volume) (Russian).

E.M. LAFULIANKE *et al.* 1991. Eurasian grassland (Russian) [M]. Peter Creux Science Press.

E.M. LAFULIANKE. Translated by Zhu Yancheng. 1959. The Grassland of Russia[M]. Science Press.

Grassland Vegetation Investigate Team of Qinghai Province. 1965. Grass Type Characteristics and its Economic Evaluation in the Southeast of Yushu in Qinghai[J]. *Plant Ecology* & *Geography Botany Datum Periodical*, 2(1).

H.H. Chiwlev. 2007. Translated by Yong Shipeng. Discussion on the Origin and Evolvement of Stipa L[M]. CAAS GRI Press (internal data).

H.Walter (translated by CAS Plant Industry Ecological Room. 1984. World Vegetation---Ecosystem of Land Biosphere[M]. Science Press.

Hao Shougang, Ma Xueping. 2000. Origin and Evolution of Life[M]. Higher Education Press, Springer Press.

Hou Xueyu. 1988. Chinese Natural Ecology Regionalization and Big Agriculture Development Tactic[M]. Science Press.

Hou Xueyu. 1988. Chinese Vegetation Geography. China Natural Geography: Plant Geography (sub-volume)[M]. Science Press.

J.M. Suttie, S.G.Reynolds and C. Batello. 2005. Grassland of the world, Plant Production and Protection Series [M]. Publishing Management Service, Information Division, FAO: Agriculture and Consumer Protection, No.34.

J.Marc Foggin. 2000. Biodiversity protection and the search for sustainability in Tibetan Plateau grasslands [D]. Arizona State University.

Jiang Shu. 1960. Upland Meadow and Forest in the west of Sichuan Province[J]. *Botany Gazette*, 9 (2).

Li Bo, *et al*. 1964. Pilot Study of moisture ecology of *Aneurolepidium chinensis* and Tussock grass Steppe community in Inner Mongolia Hulun Buir Grassland area[J]. *Plant Ecology & Geography Botany Periodical*, 2(1).

Li Bo, Yong Shipeng, Cui Haiting. 1990. The Principle, Method and Application of Ecological Zoning[J]. Plant Ecology Transaction, 1.

Li Bo, Yong Shipeng, *et al*. 1990. China C Grassland[M]. Beijing: Science Press.

Li Bo, Yong Shipeng. 1999. Cold-Temperate Grassland: Recognition and Identification[M]. Beijing: Science Press.

Li Bo. 1997. Degradation and Prevention Tactics of China's Northern Grassland[J]. *China Agriculture Science*, 30 (6).

Liu Zhongling, Wang Yifeng, Yong Shipeng, *et al*. 1985. Inner Mongolia Vegetation[M]. Beijing: Science Press.

Liu Zhongling. 1960. General Picture of Inner Mongolia Grassland vegetation[J]. Inner Mongolia University *Transaction* (*Natural Science*), 2.

Liu Zhongling. 1963. *Stipa* Formation in Inner Mongolia[J]. *Plant Ecology & Geography Botany Periodical*,1(1-2).

Ma Yuquan, Liu Zhongling, *et al*. 1995. Pilot Study of Flora in Inner Mongolia. *MaYuquan Memoir* [M]. Huhhot: Inner Mongolia People's Publishing House, .

Ma Yuquan. 1989-1995. Inner Mongolia Flora. Inner Mongolia People's Publishing House[M], Vol. I-V, (2nd Edition).

Qinghai-Tibet Plateau integration review team of CAS. 1988. Tibet Vegetation[M]. Beijing: Science Press.

Sheng Helin, Wang Peichao. 1985. An Introduction to Mammalogy[M]. Huadong Normal University Press.

Sinkiang integration review team of CAS. 1978. Sinkiang Vegetation and its Utilization[M]. Science Press.

T.II Golyev. 1957a. Vegetation Summary of Northeast China and East of Inner Mongolia. Chinese Journal of Plant Ecology[M]. Science Press, No. 11.

T.II Golyev. 1957b. An Introduction to Geobotany of Alkaline Grassland along the Railway of Binzhou in Heilongjiang Province. Chinese Journal of Plant Ecology[M]. Beijing: Science Press, No. 11.

Tibet integration review team of CAS. 1966. The Vegetation of mid-Tibet[M]. Science Press.

Wang Xiangting. 1991. Vertebrate in Gansu Province[M]. Gansu Science Press.

Wang Yifeng, Yong Shipeng, Liu Zhongling. 1979. Zonal Characteristics of Vegetation in Inner Mongolia Autonomous Region[J]. *Botany Gazette*, 21(3).

Wang Yifeng. 1991. Vegetation Resources in Loess Plateau Area and Its Reasonable Utilization[M]. Beijing: China Science and Technology Press.

Wu Zhengyi. 1983. Vegetation of China[M]. Beijing: Science Press.

Wu Zhengyi. 1987. Flora of Tibet[M]. Beijing:Science Press.

Xing Lianlian, Yang Guisheng, *et al.* 2005. Crane Resources and Protection in Inner Mongolia[M]. China's Crane Study.

Xing Lianlian. 1996. Avifauna of Ulansuhai Nur in Inner Mongolia[M]. Inner Mongolia University Press.

Xiong Yi, Li Qingkui. 1987. China's Soil[M]. Beijing:Science Press.

Xu Peng. 1982. Grassland Zoning Principles and the Discussion on Sinkiang Grassland Regionalization[J]. *Sinkiang Grassland Communication*, 3.

Xu Rigan. 2000. Inner Mongolia Fauna (Volume II) [M]. Inner Mongolia University Press.

Xu Rigan. 2007. Inner Mongolia Fauna (Volume III) [M]. Inner Mongolia University Press.

Xu Zhu. 2004. Chinese Forage Handbook[M]. Beijing: Chemical Industry Press.

Yan Chongwei, Zhao Zhengjie. 1996. Illustrated Handbook of China's Avifauna[M]. Taiwan Kingfisher Culture Co. Ltd.

Yang Guisheng, Xing Lianlian. 1998. Directory and Distribution of Vertebrate in Inner Mongolia[M]. Inner Mongolia University Press.

Yin Binggao. 1993. Rare Wildlife and its Protection in Tibet [M]. Beijing: Chinese Forestry Publishing House.

Yong Shipeng, Li Bo. 1989. Chinese Natural Protection Atlas—Chinese Grassland Distribution Chart and Its Instruction[M]. Beijing: Science Press.

Yong Shipeng, Li Bo. 1991. Problems of Grassland Ecological Monitor[J]. *Inner Mongolia Environment Protection*, 1.

Yong Shipeng, Li Bo. 1992. China Natural Disaster Atlas—Grassland Degradation (Naturalness index) [M]. Beijing: Science Press.

Yong Shipeng, Zhang Zixue, Yong Weiyi. 2001. Grassland landscape ecology types in Inner Mongolia and its failure analysis. Survey on Ecological Environmental Remote Sensing in Inner Mongolia at the end of 20th Century [M]. Huhot: Inner Mongolia Peoplegy Publishing House.

Yong Shipeng. 1985. Protective Utilization of China t Pasture Resources and Ecological Monitoring Problems (Teaching materials of operating management training class of national grassland station). Agriculture, Animal

Husbandry, Fishery & Co-operative Department [M], CAAS GRI Press.

Yong Shipeng. 1991. Grassland Vegetation. Encyclopedia of China: Geography [M]. Beijing: Encyclopedia of China Publishing House.

Yong Shipeng. 1996. Biodiversity of Temperate Grassland. Research Report of National Conditions about Biodiversity of China[M]. China Environmental Science Press.

Yong Shipeng. 2000. Grassland Patrol[J]. *Human & Biosphere*, 1:30-32.

Yuan Guoying. 1991. Vertebrate in Sinkiang [M]. Urumchi: Sinkiang People's Publishing House.

Zhang Jiacheng, Lin Zhiguang. 1985. China's Climate[M]. Shanghai:Shanghai Technology Press.

Zhang Jingwei. 1963 Basic Characteristics and Terrain Significance of Grassland in the Southeast of Qiangtang Plateau [J]. *Plant Ecology* & *Geography Botany Datum Periodical*, 1(1-2).

Zhang Xinshi. 1978. Discussion on the Plateau Zonality of Tibetan Vegetation[J]. *Botany Gazette*, 6.

Zhang Yun. 1998. Biological Evolution[M]. Beijing University Press.

Zhao Ermi, Huang Meihua. 1998. China's Fauna[M]. Science Press.

Zhao Ermi, Zhao Kentang. 1999. China's Fauna[M]. Science Press.

Zhao Ji, Chen Chuankang. 1999. The Chinese Geography[M]. Higher Education Press.

Zhao Yizhi. 1993. Taxonomical Study of China's *Caragana*[J]. *Journal of Inner Mongolia University* (*Natural Science*), 24(6).

Zheng Guangmei. 2002. Avifauna Classification and Distribution Directory of the World[M]. Science Press.

Zheng Guangmei. 2005. Avifauna Classification and Distribution Directory of China[M]. Science Press.

Zheng Huiying, Li Jiandong. 1999. Grassland vegetation and Utilization & Protection of Songnen Plain[M]. Science Press.

Zhong Yankai, Piao Shunji. 1988. Test Outcome Analysis of Mowing Succession in Grassland of *Aneurolepidium chinensis*. Grassland Ecosystem Study[M]. Science Press.

Zhu Ting cheng. 2004. Bioecology of Leymus chinensis[M]. Jilin Province Science Press.

Zhu Tingcheng. 1958. An Introduction to Dominant Grasslands in Northeast China[J]. *Northeast Normal University Science Collect Periodical*, 1.